crie, Pa 16510
739-2928

The Birth of Chrysler Corporation and Its Engineering Legacy

Carl Breer

Edited by
Anthony J. Yanik

Prepared under the auspices of the
SAE Historical Committee.

Published by:
Society of Automotive Engineers, Inc.
400 Commonwealth Drive
Warrendale, PA 15096-0001
U.S.A.
Phone: (412) 776-4841
Fax: (412) 776-5760

Library of Congress Cataloging-in-Publication Data

Breer, Carl, 1883
 The birth of Chrysler Corporation and its engineering legacy / Carl Breer ; edited by Anthony J. Yanik.
  p. cm.
 Includes index.
 ISBN 1-56091-524-2
 1. Automobiles--United States--Design and construction--History. 2. Chrysler Corporation--History. 3. Breer, Carl, 1883. 4. Zeder, Fred, 1886. 5. Skelton, Owen, 1886. 6. Automobile engineers--United States.
 I. Yanik, Anthony J. II. Title.
 TL23.B63 1994
 338.7'629222'0973--dc20   94-43487
                                                                                                   CIP

Copyright © 1995 Society of Automotive Engineers, Inc.

ISBN 1-56091-524-2

All rights reserved. Printed in the United States of America.
Second printing.

Permission to photocopy for internal or personal use, or the internal or personal use of specific clients, is granted by SAE for libraries and other users registered with the Copyright Clearance Center (CCC), provided that the base fee of $.50 per page is paid directly to CCC, 222 Rosewood Dr., Danvers, MA 01923. Special requests should be addressed to the SAE Publications Group. 1-56091-524-2/95 $.50

SAE Order No. R-144

Carl Breer: 1883-1970

Table of Contents

Acknowledgements vii

Prologue ix

Introduction xi

Carl Breer 1

Fred Zeder 5

Owen Skelton 7

Part I: Carl Breer: The Early Years 9

Part II: With Zeder and Skelton at Studebaker, 1916-1918 31

Part III: We Create Chrysler Corporation 65

Part IV: Reminiscences of Early Product Developments at
    Chrysler Corporation 91

Part V: Birth of the Airflow Car 143

Part VI: Railroad Ride Research Along Airflow Principles 177

Part VII: The Chrysler Engineering Team and the War Effort 187

Part VIII: Death of Walter Chrysler and a New Regime 197

Conclusion 199

Epilogue 203

Index 213

Acknowledgements

This book exists primarily because Dr. James Flink brought the existence of Carl Breer's manuscript to the attention of James Wren, then head of the Society of Automotive Engineers Historical Committee. Mr. Wren contacted William Zeder Breer, son of Mr. Breer, Sr., with the ultimate result that SAE purchased the manuscript which I agreed to edit for publication.

I am indebted to Mr. Wren for his support through the several iterations of this editing task, and especially to Mr. William Zeder Breer for his assistance in making certain that any changes I made accurately reflected his father's thoughts. Mr. Breer also was kind enough to make the photographs available from his father's many scrapbooks for copy in this book, a number of which have not appeared before in other historical publications.

Thanks also to Karen Prymak of the American Automobile Manufacturers Association who made sense out of my numerous penciled changes, sentence and page shifts to type out drafts that finally resulted in the text that makes up this book.

Anthony J. Yanik

Prologue

This is the story of a remarkable engineering team as told by one of its members, Carl Breer. He together with Fred Zeder and Owen Skelton brought solid engineering principles to the design and testing of early automobiles at a time when such principles were still in the formative stage.

Beginning with a look back at Carl Breer's early years, the book's focus then shifts to the Zeder, Skelton, and Breer engineering team. Through the eyes of Carl Breer we are given a glimpse of the trio as they provide Studebaker with a solid engineering base prior to World War I and, more importantly, become the building blocks upon which the present Chrysler Corporation was founded. Mr. Breer then takes us on a tour of the multitudinous innovations for which the trio were responsible during the early days of Chrysler, capped by a behind-the-scenes description of the company's most well-known make — the Airflow — an engineering tour de force but marketing failure. Unknown to most are the numerous applications of Airflow principles to other Chrysler products, including the search for a better riding railroad train. Finally, Mr. Breer revisits the various efforts that Chrysler pursued for the U. S. government during World War II.

While Mr. Breer did not date his manuscript, it appears that he wrote it over an extended period of time, completing it in 1960.

I have attempted to adhere to his original words as closely as possible, if not in words then at least in thought in places where his original language became obscure over time.

It is only appropriate that brief biographies of the three principles involved — Carl Breer, Fred Zeder, and Owen Skelton — be included to flesh out those years prior to World War I before they came together to form one of the most productive engineering teams in automotive history. The biographies are provided by Mr. Breer himself.

Anthony J.Yanik

Introduction

This is the history of three engineers who joined together to form a business association during the early days of the automobile, an engineering team that was able to hold together for 35 years of productive effort. The team: myself (Carl Breer), Fred M. Zeder, and Owen Skelton. Looking back, we three helped lead the way in popularizing many basic new innovations that have helped prepare the solid foundation on which our great American automotive free enterprise system has been built.

Our primary goal was to recognize and strengthen each weak link that we could uncover in the automobile chain of mechanically related things, cure them, and make them more efficient yet free from human tampering. The end result, the modern automobile of today, is one that is more dependable and safer than the people who drive them.

It was indeed a fortunate occasion when that opportunity opened up in 1916 at Studebaker for the three of us to come together, recognize how each of our strengths helped to complement each other as an automotive engineering team, and have the perseverance to become a vital factor in the development of the product.

Later, it became the lab of our engineering team to forge the engineering foundation that made it possible for Walter Chrysler to found the present Chrysler Corporation which provided a livelihood for hundreds of thousands of people along with countless others associated with supplying industries.

Most of the automobile innovations we initiated and commercialized have been accepted as standard equipment on all makes of automobiles today, such as hydraulic brakes, floating power, downdraft carburetors, air cleaners, oil filters, curved windshields, and many others — the evolution and development of which are recorded in the following chapters.

Readers who have lived during our time can appreciate the phenomenal, rapid change in our living environment during the period of transition from the horse and buggy and dirt road to the modern motor car and highway. Physical work has been greatly reduced by mechanization, both in the plant as well as in the home. Heat now is delivered to the user automatically.

During my youth, for example, all work was accomplished by hand, or foot. To illustrate, at the turn of the century I built myself a complete steam car using Barnes foot-operated lathe and hand tools. We have come a long way since then as the following pages will indicate, most surprisingly within one man's lifetime — mine.

When I was growing up, we got from here to there by harnessing a horse to a vehicle, using reins to steer and a whip to accelerate if hoping to make time to reach one's destination on schedule. On long trips it was necessary to carry feed for fuel, and to take time out for both man and horse to rest after jouncing up and down over the rough roads of the day.

The sense of freedom that came from one's first ride in a horseless carriage, going faster (and bouncing higher over those same roads), and buying fuel from grocery stores in five gallon tin cans was a new thrill. The physical strain in the horseless carriage was much higher over those rough roads, but the exhilaration of being your own master was well worth it in spite of the fact that a horse and buggy gave one a more comfortable ride.

Also in those days came the interest — as well as the need — to become interested in learning about the mechanical things under the hood. Service stations were few, the best solution being to dig out the instruction book. Soon you become your own mechanic, a far cry from grooming and hitching a horse.

Gradually, as cars became more tinker-proof, we lost interest in doing our own maintenance work, coming to the conclusion that automobiles were becoming pretty reliable products. Now we began to look for features adapted to more personal comfort, or gave a car a more distinctive appearance, namely its styling. Appearance, however, was not much of an overloading factor when Fred Zeder, Owen Skelton, and I began our association in 1916 at Studebaker. Durability and reliability were the challenge with which we were initially faced.

Fred Zeder's association with the automobile industry came before Skelton's and mine. He could recall the days when many auto suppliers entertained lavishly, and how many a contract was settled over the old Pontchartrain bar. The Pontchartrain Hotel has long since disappeared. In later years, the auto suppliers adopted the habit of lavishing Christmas presents on us until they reached the point that corporations had to issue orders for vendors to cease.

Our engineering trio discouraged this sort of outside influence and made our feelings clear to our employees, especially the new young engineers coming in. All of our development work had to be based on sound engineering principles. Actually nature's fundamental laws are all related and none are contrary or conflict with one another. Man himself may be contrary or arbitrary at times, but Nature's laws have no choice. There's only one answer and that's the right answer. Men that were not honest in their profession did not last long in our organization. Nature's laws are many and often difficult to interpret, but as we explored and discovered them, so with respect to them did we build our products.

In looking back, I can see that the three of us were fortunate to come from three entirely different backgrounds during an important, evolutionary era of the motor vehicle industry. Although this book is primarily about our 35-year association, I feel it necessary to relate my own early years in order to give you some insight into how automobile engineers came to be in those pioneering days when such an engineering curriculum scarcely existed on college campuses.

The four horsemen who created Chrysler Corporation:
(top) Owen Skelton and Carl Breer; (bottom) Walter Chrysler and Fred Zeder.
(Breer Collection)

Carl Breer

Carl Breer was born in Los Angeles, California, November 8, 1883. He acquired craftsmanship experience in his father's blacksmith shop and at an early age designed and built his own steam car. With the ultimate goal of enrolling in Stanford University, he drove the steam car to Pasadena, and demonstrated it to the Stanford-accredited Throop Polytechnic Institute. (Throop later became the highly-regarded California Institute of Technology.) Throop accepted him for a one-year term (1904), and graduated him with full credit requirements that enabled him to enter Stanford University.

Building and demonstrating his steam car to members of the Tourist Automobile Company (the only west coast manufacturer of automobiles) brought him his first job, one of several automotive positions involved in demonstrating and servicing such eastern makes as Toledo Steam Cars, The Spalding, The Northern, and The White Steamer. He also was to become associated in the design and building of the Duro Car Company in Los Angeles, manufacturers of a two-cylinder opposed engine and transmission set forward with a cardinal shaft (propeller shaft) drive to the rear axle.

After graduating as a Mechanical Engineer from Stanford in 1909, Breer received an invitation to join Allis-Chalmers. The Allis-Chalmers Manufacturing Company, located at West Allis, a suburb of Milwaukee, had initiated an attractive two-year apprenticeship course selecting mechanical engineering graduates from many top universities. Some twenty-five were chosen. Breer was one, as was Fred Zeder, a graduate from the University of Michigan. It was at the Allis-Chalmers Company at West Allis where Fred Zeder and he became close friends.

In 1911 Breer returned to the west coast to become superintendent of the Moreland Motor Truck Company building factory. Still on the move, he went into the service and accessory business, organizing the Home Electric Auto Works in 1914. He soon became disenchanted with this effort and sold out to his partner. In 1916, just as Breer had completed a combination experimental garage and shop, a letter arrived from Fred Zeder asking him to join Studebaker (of which he had become chief engineer) to set up and organize a research division.

Breer as he completed his apprenticeship at Allis-Chalmers in 1911 and prepared to return to California. (Breer Collection)

Carl Breer (bottom left) and Fred Zeder (top left) pose with their fellow apprentices at Allis-Chalmers. The two young engineers formed a friendship that extended throughout their business and social lives, and included the marriage of Breer to Zeder's sister. (Breer Collection)

Fred Zeder

Fred Zeder's birthplace was Bay City, Michigan, where he was born on March 19, 1886. He graduated from the University of Michigan in 1909 with a Bachelor of Science degree in Mechanical Engineering. Immediately after graduating, Zeder was selected to the apprenticeship course of the Allis-Chalmers Company. After leaving West Allis, he completed a power plant erecting job in Detroit in 1910, then later that year joined the Everett-Metzger-Flanders Company (E.M.F.) where he assumed charge of its laboratory while still acting as a consultant in power plant work.

E.M.F. was located in the center of Detroit's many automotive activities on Piquette Street east of Woodward. Nearby, Fisher Brothers built auto bodies for E.M.F. Also on Piquette Street were the Regal Car Company and the Henry Ford Company. General Motors at that time had a laboratory near Woodward and Piquette consisting of a 50 x 100 ft two-story building with a foundry below.

Then, in sequence, the Studebaker Brothers of South Bend took over the E.M.F. enterprise and started to build their own car bodies. In 1913 Fred became consulting engineer for the Studebaker Corporation, and a year later was promoted to chief engineer. The production of Studebaker cars reached a peak in 1916, but accumulating car troubles and field problems depleted the company's finances, heading it toward possible receivership. It was at this time that Fred Zeder, realizing the weaknesses of his engineering organization, decided to bring in Owen Skelton and Carl Breer to form a team with him to take a new engineering approach within a new organizational setup.

Owen Skelton

Owen R. Skelton was born on February 9, 1886, in Edgerton, Ohio. From 1905 to 1907 Skelton worked at the then great Pope-Toledo plant which in turn became the Willys-Overland factory headquartered in Toledo, Ohio. Pope-Toledo was the most progressive of all gasoline motor cars thanks to its advanced engineering.

Skelton's natural instinct and professional training led him to the design drafting department of the Packard Motor Car Company in Detroit where he became a transmission specialist. His experience with Packard proved invaluable to his future, joining Zeder in 1914. Skelton's design background, especially with rear axles, drives, and gear boxes, proved extremely helpful in expediting the design of Studebaker's new line that debuted in 1918.

Part I:
Carl Breer - The Early Years

My father, the youngest in his family, was born in the Hartz Mountains in Germany in 1828. His trade was blacksmithing. He traveled from village to village as a journeyman to add to his skill. Just before he turned 21, he decided to come to America, primarily to avoid being conscripted into the army for three years which he felt would be a total waste of his time. He worked his way across the Atlantic on a sidewheeler repairing anchor chains. After landing in Hoboken, he visited an aunt in Indianapolis, Indiana, went on then to New Orleans, across Panama, and boarded a ship to San Francisco. He took one look at the placer gold mining operation and did not like it, so he boarded the same boat he had come in on which took him to San Pedro and never left Los Angeles the rest of his 80 years.

Mother was born in the Black Forest in Germany in a little farming community at the end of the road known as Ober-Owerisheim. (It was still a quaint little village when I visited it in 1937 — a small running creek in the center of a main dirt thoroughfare, little foot bridges and village pumps adjacent to the stream for the dwellers nearby who came out in wooden shoes.) My mother and her younger sister came to America to join an uncle who was operating coastal vessels along the Pacific coast. They, too, settled in Los Angeles where they spent the rest of their lives. My father and mother married in 1863.

My father worked for John Goller who operated a blacksmith and carriage shop. He had three assistants who were named Louis — Roeder, Lichtenberger, and Breer. Of the three, my father was known as "Iron Louie" because of his magnificent physique. He worked by his anvil for sixty years and died of old age on his 81st birthday, August 25, 1909. Most of the time he operated his own shop.

Our family first lived in an adobe at 215 South San Pedro Street. Later my father built one of the first brick houses in Los Angeles. It was just north of the adobe, leaving an entrance driveway between. Most of the Breer children were born in this brick house, I being the youngest.

On one side to the rear were the stables for the horse and cow and hayloft overhead. Added to this were pigeon lofts and coverage for the family surrey buggy and tools to curry the rear garden.

As the family grew, additional outside entrance rooms were added immediately behind the main house with a utility porch between where the family washing was done. The porch was open and a narrow portion extended clear across the rear of the main house. A honeysuckle vine closed it in and shaded the afternoon sun. The new addition was up to date in that it had a separate bathroom, a modern zinc-lined bathtub, and a washstand with running water. In those days the toilet was an outdoor "Chick Sales." At the north open end of the porch through which the breezes blew was hanging the family Olla, a porous earthen jar with a wetted burlap covering, the evaporation of which really kept the porous container full of drinking water cool.

At the rear of our property were apple, peach, pear, apricot, fig, and orange trees and at the northwest rear corner under a big willow tree was an earth-lined swimming pool. Of course, we had a vegetable garden fenced off to keep the chickens and rabbits away.

Immediately behind the adobe, not far from the street, dad had his large, open-front workshop. In those days there were no power tools so everything was done by hand.

My father could work wonders with steel or iron, whether making a plow share or an iron tire for a wagon wheel, yet he never shod a horse.

Growing Up in Los Angeles

The more I think back, the more I realize how fortunate I was to be born at the trailer end of a thrifty family. I had older brothers and two sisters who fulfilled every desire, whether it came the latest in tricycles, bicycles, or Kodak cameras, as well as vehicles we made ourselves such as special coasters and dog-drawn wagons with leather to make our own dog harness.

As far back as I can remember, mother and the family had places at the Pacific Ocean seashore. My father and the older boys would came down weekends by horse and buggy. In the spring as summer came on, five or six families would move down to Santa Monica Canyon onto the ocean sand where a beautiful stream of canyon water winding its way through green sycamores ran into the ocean. It really was a beautiful spot. From our home, the long 16-mile trip through the green countryside to the shore took about three and a half hours via horse and buggy. Occasionally a round trip was made on the same day which was a wearing trip for both horse and driver. Later we spent other summers in Santa Monica where we pioneered building a redwood cottage on the sand with permission of the property owners who were friends.

On the more mundane level, my usual day involved going to school, after which there was wood to chop and bring in for the kitchen stove, to say nothing of pitching in as father's and brothers' helper, swinging the ten- and later twelve-pound sledge, helping to weld or put new cutting edges and points on old and worn plow shares, or welding and shaping new heavy metal wagon wheel tires. We would build fires around the wheel tires to expand them so that we could slip them over the wood rim wheels, then shrink them tight with cooling streams of water so as not to burn and ruin the wood rim.

It was a great experience for me to watch the heating forge and see how two pieces of steel or iron could be welded into one, to learn why borax iron fillings were added, or why the temperature of the sparking iron had to be just so. This knowledge of working with metal and various hard wood timber proved to be invaluable to me later in my professional life.

Roots of My Interest in Engineering

Before finishing grammar school a very important incident happened in my life. My brother-in-law's brother, Gustave Schroeder, brother Bill, and I took a stroll one beautiful spring Saturday up through Elysian Park to view the city from the high hills over the Los Angeles River valley.

Before long, we came upon the Los Angeles city water works pumping plant. As we stood looking in, a very kindly gentleman, Mr. Fred J. Fisher, then Chief Engineer of the one and only Los Angeles Water Works, greeted us and invited us on a tour of the plant. As we walked around, my eye caught notice of a bi-polar direct current generator, sitting on a high bracket on the wall. It was operated by overhead leather belts running from a counter shaft driven by a Pelton water wheel. I was thrilled to learn that Mr. Fisher had personally built this electric generator in this very plant, set it up, and used the force of the water from the reservoir above to operate it so that he could light up all the dark corners in this remote pumping plant both day and night.

Fisher's plant was filled with engineering innovations. For example, another Pelton water wheel operated Fisher's stand-by engine lathe which he used when some minor repair parts were quickly needed. What a unique setup! When things were running smoothly Mr. Fisher had spare time for his own use.

Before I left I was itching to build a machine of my own, one that could make electricity to light up an electric light bulb. From then on Fred Fisher became a life-long friend, and I was befriended like an adopted son.

After school hours and during weekends I worked on my homemade generator. On my bike I shuttled some five miles to Fisher's plant to become educated in operating his lathe to make the parts I needed. What a thrill it was later to see electric bulbs glow from electricity generated by the machine I built.

Fred J. Fisher, Chief Engineer, Los Angeles Water Works, at the turn of the century. Fisher was Breer's early engineering mentor. (Breer Collection)

Inspired by Fisher's engineering innovations, Breer, at age 14, built his own generator (shown on the left) which he used to light electric bulbs in the Breer homestead. On the right is an electric motor that he later constructed. (Breer Collection)

Building My Own Steam Car: 1900

One day, I confided in Fred Fisher my desire to build a motor-driven car. Fisher, always open minded, asked me whether I would be better off attempting to build a gasoline engine, which we know nothing about, or a steam engine we know more about. So steam engine it was. I found a description of the Stanley Steamer's working parts in a magazine which guided me at the time.

Brother Bill was also interested. He had some money, so we plunged in and bought a Barnes ten-inch foot-powered engine lathe. We installed it in the old adobe homestead which was old and vacated. I worked day and night building a work bench and cutting through the shingle roof to install a skylight. I wired the house for lights too.

Then I made a dimension drawing of a two-cylinder, double-acting steam engine with Stevenson link motion to operate the valves for forward and reverse. From this general drawing I made details for each part as I needed it. I carved patterns of sugar pine and poplar for the foundry castings, starting with the most difficult parts to cast first like the cylinder block and valve chamber. I took the patterns to the foundry, but after several failures they gave up on casting the cylinder block because of difficulty with the delicate steam passage cores. I asked if I could give it a try. I made the cores, set the mold, and luckily was successful in getting a nice satisfactory cylinder casting. This was my first big obstacle to overcome. Brother Bill helped me forge the steel crankshaft, the connecting rods shapes, etc. These, along with the bronze castings for bearings and miscellaneous parts, were then machined.

Proud was I the day with horse and buckboard I took the completed engine to Fred Fisher's plant in Elysian Park to check it out and set the valve gear in operation. I then located the desired tubular running gear with wheels and tires which were being made for steam cars in Buffalo at the time, and ordered them. The 16-in one-piece, drawn steel head and shell boiler was also purchased.

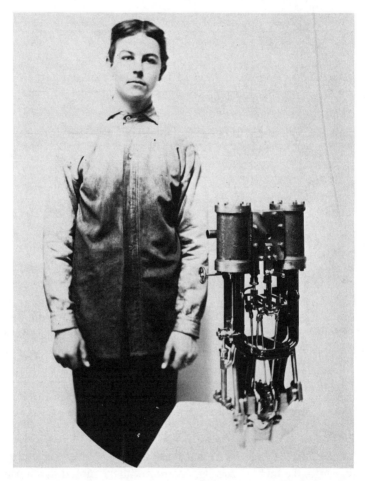

Breer with his homemade two cylinder, double-acting steam engine. When the local foundry gave up on casting the cylinder block to his design, Breer made the cores, set the mold, and was successful on his first cast. (Breer Collection)

Carl Breer - The Early Years

Since I was using gasoline to heat the boiler, I designed patterns for a unique one-piece, cast iron, multiple Bunsen type of burner. This consisted of a disc with an upper and lower sheet cored out between through which were drilled air tubes. A number of small holes then had to be drilled into the inner chamber around each air tube to act like Bunsen burners, close to 3,000 of them. No practical power source for drilling these was available at the time, so I designed a high-speed Pelton waterwheel type of direct drive drill press to operate on water pressure from our outdoor water faucet. It worked well, and in a short time the several thousand holes were drilled.

I also decided that the standard method for starting a burner was too cumbersome. It consisted of heating a "U" tube over an outside fire, then quickly inserting and attaching it so that a small flow of gasoline vapor would ignite, and by careful manipulation, provide sufficient pilot burner heat to start the main burner.

I simply added an extra volume air tank. By means of a mixing valve, I carbureted the raw cold gasoline into a burning mixture to operate the main burner directly long enough to maintain its own gasoline vapor supply. As time went on, I designed, framed, and built the body using poplar wood for the exterior and ash for frame structure. Carriage workers trimmed the upholstering and painted the body for me.

Finally, the day arrived to see if this masterpiece would run. It was a late fall afternoon in 1901, in the yard by the blacksmith shop, that I started up the burner. It worked. Pressure showed on the steam gage, and the water level in the water level gage bobbed up and down to indicate activity within.

I nervously got up in the seat and pushed on the throttle. The car suddenly leaped forward in action. I tried it in forward and then in reverse. Each time I started to open the throttle, the car jumped forward. It was too late in the afternoon for me to take it out on the street, and I needed to tone down its instant jump start. I turned to a conical valve stem to graduate the throttle valve opening which tamed it down. On this next attempt when I dared to take the car out on the street, what a thrill it was to drive along quietly, without hoof clatter. And it was a rare privilege since not many people could say that the first car they ever drove was one they built. I was proud of my achievement!

Now I was anxious to have Mr. Fred Fisher join me in the thrill of riding this mechanically-animated thing. The following Sunday, Mr. Fisher arrived on his bike. After driving him around, I jumped out and asked him to take over. At the time, San Pedro Street was a well-crowned, fairly narrow dirt road with rather

Young Breer (he was only 18 at the time) poses at the throttle of his completed steam car. A school chum, Fred Ward, is in the adjacent seat. (Breer Collection)

The Breer steam car had its maiden voyage in the fall of 1901. The only work Breer did not do himself was to paint the body and trim the upholstery. It is shown here as restored by Gilbert Goode at Chrysler Research some 40 years later. (Breer Collection)

Carl Breer - The Early Years

The Breer steam engine in place. To drill the three thousand small holes into the inner chamber needed around the air tubes, Breer devised a drill press that would operate off of water pressure from his outdoor water faucet. (Breer Collection)

deep drain gutters on either side. While attempting to turn around, Mr. Fisher became confused at the controls and into the ditch we went, bending the right front axle. I never forgot the expression on his face. He said, "I never wanted that to happen," and again "I would not have had that happen for anything." My heart went out to him.

Since we were only a block from home, we got a plank and some blocking to keep the misaligned wheel off the ground, and harnessed up our old family dobbin, "Fanny." It was not long before we managed to drag the vehicle home. Before another weekend, everything was as good as new.

Over the next few years I made various refinements: The cylinders were jacketed with iron and dressed with brass fastening bands; a new and unique independent steam operated water pump was designed by myself having a trip-trigger valve mechanism to make the single, double-acting steam pump continue to work without stalling. I added a steam whistle, foot-operated, and for long distance travel, an auxiliary gasoline tank made of cast brass heads. This was mounted on special forged steel brackets extending from the rear of the body. This was completed with difficulty about 4:00 one morning so that brother Bill and I could go trout fishing as planned at the mouth of the San Gabriel Canyon at Azusa, so we left at 6:00 that same morning. It was beautiful spring day and we caught plenty of trout. It gave me a lot of satisfaction and delight just to think we were able to drive some 35 miles over dirt roads and return and still have plenty of time to wade the stream and flycast for fish. This wouldn't have been possible by horse or buggy.

I might explain that gasoline then was a by-product of kerosene and could only be bought at a few grocery stores. It was sold like domestic kerosene in five-gallon, rectangular tin cans. For safe handling they were packed two in a wooden box, and being an excess product of kerosene, sold for less. A five-gallon, non-returnable can of gasoline sold for about 90 cents, while kerosene, being less dangerous for stoves, and also being used for lamps, sold for $1.25 or $1.35 a can. At that time Los Angeles had just demanded that horseless carriages be registered and carry a city license number plate. One had to go to the city hall to register the number desired. I requested 999, but someone registered that number choice ahead of me. I compromised on another, a favored locomotive number, 666. The license was made of patent leather with metal numbers attached, and was conspicuously hung in front with leather straps from the running gear.

The operation of a steam car is very simple as compared to that of the early gasoline car. There are no gears to shift, and no clutch to operate. All is direct drive, using the throttle lever for speed and power. There is a second lever for backing. A simple two-cylinder steam engine is double acting, providing the same number of force efforts for each engine revolution as an eight-cylinder gasoline automobile engine. Pressure from the boiler is carried on throughout each stroke greatly increasing the vehicle's ability for hill climbing without the need to change gear ratio as required with a gas engine. Nevertheless, the gasoline engine eventually replaced the steam car for obvious reasons: It starts instantly, is more efficient in miles per gallon, is relatively compact, and has extremely high power ability.

Rollin White, for whom I did consulting work while developing the Rollin gasoline car, stated that while still producing their well-known White Steamer, he and his brother made a trip to Europe and consulted with the best steam authorities in England. They came back confirming the English engineers' conclusion that steam could not compete with a gasoline car under 1,000 hp rating. For this reason, he subsequently changed the White Company's steam vehicles to the gasoline engine type.

People have asked as to the rated speed of my own steam car. I would have to say that it was twice horse speed, or 25-35 mph, especially good for the rough wagon roads of the day. At that, it was much quicker for distance travel compared to a horse because it eliminated animal fatigue. And for hill climbing, there was plenty of added ability.

Sometimes a trip in my steamer became a major experience. On one occasion, while attending Throop Polytechnic School, Walter Vail (son of the then famous Walter Vail, Sr., Vail & Vickers Cattle Ranches) and I, dressed in tuxedos and stiff shirts, left our house in Los Angeles and headed to Pasadena. I left Walter off at the corner of Colorado Street to take the Pacific Electric to the Pintaresca Hotel for our school dance. I then drove to Orange Grove Avenue where I picked up my fair lady, Jennie McLain. I could not take Walter and his date with us because the seat had room only for two. We then drove to the hotel. After midnight, the party broke up and I motored Jennie home. With thanks for the success of a grand evening and a brief "good night" I returned to pick up Walter and take him home.

Just as I bumped over the Union Pacific Railroad tracks that crossed Colorado Street something gave way. I stopped and discovered that I could only drive in reverse, therefore I had to proceed to our meeting place driving backwards. Here Walter happily greeted me thinking that I was just showing off on my return to pick him up! I told Walter that I seemed to have picked up something that cut off my forward-going valve gear. We certainly didn't want to go to L.A. backwards, yet we didn't want to leave the car where it was. Eventually I found a stable whose keeper gave me a suit of overalls. A quick exploration indicated that I had picked up a length of bailing wire which had caught on the chain, wrapped itself around the engine shaft, and broken off the two adjacent eccentric rods operating the forward-going valve gear. After several hours, I was able to transfer the reverse eccentric rods to the forward-going eccentrics but then found that the eccentric blocks had shifted around the engine shaft, upsetting the timing. Finally, I was able to make guess settings of the eccentrics that provided forward motion in a galloping action. It was close to daybreak when we started on our way back.

More than once we had to get out and push. Later when I saw Walter he said he had climbed up the outer porch to the second floor and got into his room without awakening the folks as this was on Sunday morning. Likewise no one knew what time I got in at our home.

There were many other such adventures that I experienced in my steam car.

My First Job: The Tourist Automobile Company

The Tourist Automobile Company was founded in Los Angeles in 1902 through the efforts of a Mr. Hanson who built the sample car around which the manufacturing company was organized. There were quite a few "Tourist" cars produced during the company's eight-year history. Maximum output later reached about two or three cars a day. The basic design was a two-cylinder opposed engine located amidships under the seat with planetary transmission and chain drive, wood frame armored for strength with side plates of steel, and full elliptic springs at all four suspension points, plus adjustable struts for the chain drive to the rear axle. A novel additional feature was an outside lever control which clamped pairs of ash or hickory wood blocks against the low forward and reverse speed planetary drums.

Wanting experience and with thoughts in mind of going to college, I drove to the Tourist Auto Plant, then located on North Main and Alameda Streets, and I contacted Mr. Ford, head of their service garage. I told him that I had never worked for anyone before but did have as a reference my steam car outside. After looking over my product and taking a short spin he said, "You can come to work tomorrow," much to my surprise. So I had my first job. It was at 30 cents an hour, fairly high pay at the time. I was associated with the Tourist Company for several years thereafter during summer vacations working in service, assembly, and design.

My One Year at Throop Polytechnic

This manufacturing experience stimulated my desire for a college education. My brother-in-law's brother, having attended Stanford, highly recommended that university for its engineering. At that time it was a highly rated non-tuition institution, but it was rather difficult to get into.

After reading over the Stanford catalogs I was enthusiastic to attend, but I discovered to my dismay that my two and a half year's completion of the then three-year standard course at Los Angeles Commercial High School had no credit standing whatsoever with regard to entering the Stanford engineering course. Upon reading the text, I gathered that I would have to take a four-year course involving accredited high school science courses before I could even take an entry exam.

I could not see myself spending four more years of preparatory work to enter college. Fortunately, the Throop Polytechnic Institute, the forerunner of Cal-Tech, an established private school with an outstanding reputation, became my entry card.

One fine fall Saturday afternoon my commercial school associate, Chester Neiswender, and I drove to Pasadena in my steam chariot, looked around the Throop laboratories, and contacted a Dr. Perkins who headed the engineering, design, and mathematics division. After discussing my desire to enter Stanford as soon as possible without spending another four years of preparatory work, we went for a spin in my car, driving some five or ten miles. While we were gone, Mr. Neiswender talked to a Mr. Ford who had charge of the shop work courses. When we returned, Mr. Ford looked over my handiwork, and after a brief discussion, stated it would be wasting my time to take his shop training course.

I then asked Dr. Perkins how long it would take me in attending Throop to obtain the necessary credentials to enter Stanford. Dr. Perkins surprised me by saying, "I think we can put you through in a year's time." He then went to check with Throop's president, Mr. W. A. Edwards, and Mr. Edwards approved.

I started in mid-winter and finished my year at the end of the fall term of 1904. This allowed me to take the Stanford entrance exams and start the mechanical engineering course headed by Dr. W. F. Durand in the fall of 1905.

At Stanford University and the Great Earthquake

Having fulfilled the academic requirements for entrance admission, I lived in the Encina dormitory for a period of two years and was rushed by several fraternities. At the end of the first year the fraternity I later joined pressured me with the ultimatum of now or never. Desiring to stay in the Encina dormitory two years, I stalled for I felt certain that the Encina environment had much to offer. Living in Encina I was fortunate in having a senior student, Haines W. Reed, for a roommate. We had an upper fourth floor west corner room on the courtyard side from which we could view the westerly half of the Stanford Quad.

Just at daybreak, April 18, 1906, I was suddenly awakened by a tremendous shaking, crashing, and low rumbling noise! My immediate impression was that mother earth was struck by one of our planets, shattering our earth into pieces and suddenly we were traveling through space. Without warning the end had come!

I found that the heavy thick plaster from the ceiling and the sidewall had broken loose and fallen on top of me and my bed. With blinking eyes I looked across the room and I could see our hanging center light swinging to and fro. The bookcases on the wall toppled over from the vibration. Added to this was an unforgettable weird very low pitched rumbling noise — like distant rolling thunder only of a much lower pitch — a sound I had never heard or realized possible before. I could also hear the voice of my roommate, Haines Reed, who was on the other side of our room under his bed. "Earthquake! Earthquake!" he shouted. Having spent several years with his father and mother mining in the earthquake region of the Balsas Mountain several days horseback ride southward of Mexico City, he knew what earthquakes were all about.

Out we ran in bare feet and pajamas, over plaster and glass strewn on the hall floors, down four flights of stairs and out the rear door of our wing. Luckily we got out without injury. Large blocks of sandstone were laying all around us and we could see heavy keystones that had been shaken out from above the windows. The heavy roof structure loaded with its red tile had shuffled about and forced the top stone layers to fall to the ground. There was a sheer open corner crack several inches wide starting above our room and coming all the way down.

When we later returned to our room, we discovered that it was a mess. The wall on the court side had moved out several inches. Another inch, and our floor would have fallen to the room below. Daylight came in through the wide crack in the corner. The keystones over our windows had fallen out. Suddenly our room began to rock again. Along with this came that weird low pitched warning sound of the rumbling earth. It was a gruesome feeling one never forgets — that earthwave beat of extreme low rhythm.

Again in a fearsome rush, I grabbed my derby hat and overcoat and ran down in my bedroom slippers. Haines was dressed about the same. We then decided to get to the drug store for supplies to help others. On the way we stopped at the fire house and grabbed a fire ax to help break in, but never needed it. All the glass in the doors and windows of the drug store were shattered. Everything had been knocked off the shelves and was heaped on the floor.

Cautiously we entered and selected bandages and medication and then hustled back to Encina. By this time doctors were on the scene. Fellow students were at work rescuing six boys who had been caught in the three rooms over the office, one above the other. We discovered that one was killed when he saw his roommate jump out of bed as the shake started, and was on his way across the room toward the front window to see what was happening when the chimney top came through the ceiling and crumpled him in the middle of the floor. Fortunately each of the other five were trapped in their beds sandwiched between each floor with the miscellaneous furniture keeping enough space between. It took some five hours before these boys were cleared, some with broken limbs and bruised backs. How fortunate it was that not one fire started in any of the campus buildings.

With the university closed, Ralph Hopkins and I rode our bikes to San Francisco that very afternoon to stay overnight and see if we could be of any help to friends in the city.

Soon we came face-to-face with a shower of cinders from San Francisco which blinded us. To keep the cinders out of our eyes we bought a veil for each of us and trimmed them around our hats. Then we came to a tremendous gushing of water, the result of the earthquake shearing the two large water mains, the one and only water supply for San Francisco which came from the Crystal Lakes in the hills behind us. This was the basic cause for the disastrous uncontrollable San Francisco fire — no water. We were told later by the watchman at the Stanford residence across from the Fairmont Hotel that one bucket of water would have saved it from burning. Fire fighting engines and apparatus later were seen on the streets in the downtown area, masses of twisted iron. The city now is protected with many sources of fire fighting supplies of water from reservoirs below street level strategically located through the city.

Looking down into San Francisco proper immediately following the 1906 earthquake. The quake had sheared the water mains which had made the fire fighting task all but impossible. (Breer Collection)
Note: All of these earthquake photographs were actually taken by Carl Breer himself using his folding bellows-type Kodak camera.

As we came over the edge of the south San Francisco hills we could plainly see the tremendous fires in the distance, pulling the air into the burning edges, and with tremendous updraft force caused by the heat, sending smoke and debris cyclone-like skyward. The center of the circular burning area acted like a huge chimney. Where frame houses once stood there were only brick chimneys and street poles. There were no signs of ash around. It was simply lifted up and carried away.

As we came closer to the fire area we realized everyone was on the move. Anything that had wheels was a conveyance. Horses and wagons were at a premium. Bedsteads that had rollers were conveyances. Trunks dragged on casters saved belongings of people leaving homes in the fire zone. We worked our way toward the Pan Handle of Golden Gate Park where my cousins, the Rudolph Volmers, lived but they were too loaded with friends to give us a place to sleep. They did offer a horse and open-type express wagon that belonged to some of their friends who had moved out of the fire zone. We drove it out to a large open section of Golden Gate Park where we could tie the horse to the wagon, and sleep flat on the floor with our heads under the seat to keep the cinders and debris from falling on our faces.

Beginnings of a tent city in the Mission Park area of citizens who had been burned out of their homes. (Breer Collection)

Early before regular working hours we took back the horse and wagon, picked up our bikes, and worked our way as close to the fire zone as we could. The city was quickly organized and taken over by the soldiers from the Presidio under Marshall Law. Everything was out of commission — no electric power or transportation. All means of communication were cut off. There was chaos and confusion everywhere. People seemed to be going to and fro aimlessly, carrying what they could or pulling them on anything that would roll or slide.

On one of the streets crowded with vehicles and pedestrians we saw an excited team of horses drawing a covered wagon with a family and their belongings. The wagon pole rammed a single-seated surrey on which was seated a large, husky, burly fellow with a cowboy hat sitting erect, pompous-like. His single horse drawing this trim rig felt the force of the ram and began to leap and bound. Seeing what was happening, I instinctively rushed out and grabbed the horse by the bridle and while I had control of the horse, the fellow on the seat pulled out a huge six shooter, and looked back at the two fellows who had struck him. I myself jockeyed around in front of the horse in case the gun began to explode. The two fellows at the rear completely unharnessed their horses before they dared to back out the wagon and pole.

Once removed, things quieted down. The large, burly elderly individual jumped out of his rig, put his gun in his holster and came forward to me. Without a smile he said, "You are a Stanford man, are you not?" Well, I, in surprise, was happy to admit that and he proceeded to say, "I'm a Stanford man, too, from the Class of '93 [or '94.]" I don't remember which. Anyway, in the excitement I was sorry I did not get his name.

The fire was moving fast in all directions away from the downtown section, the furnace center of the previous day. Along the waterfront, there were fire fighting tugs, but they were of no assistance to the city interior. The flames traveled northwesterly, westward, and southerly stopping at wide Van Ness on the westerly edge. Out of curiosity we went to the highland of the Nob Hill area where we could look easterly into the encroaching burning area. We arrived below the Fairmont Hotel where soldiers forced us around California and Mason Streets. We took pictures of burning Chinatown in the hollow and watched soldiers dynamite houses and structures ahead of the fire trying their best to stem the fast-burning tide. At times it just seemed as if the flames reached out and set afire the exploding debris before it reached the ground. People set their own homes on fire by starting to cook without realizing that their chimneys had been shuffled out of place. Soldiers soon patrolled the areas with proclamations warning and arresting anyone who attempted to cook indoors or had a light of any kind burning in their house. When the fire burned out the city was in complete darkness. As we moved along we could see fires burning all around the Fairmont Hotel, a stone structure which we later found was the only building on Nob Hill that was not gutted by flame.

With no place to stay we worked our way south across the city taking a route not yet burned, nearer to Van Ness Avenue. It was getting dark by the time we stopped on another easterly slope of a large vacant hill wedged between two streets leading westward from upper Market Street. Ralph and I had an excellent view from this point. This area was completely filled with people and their belongings all set to shelter for the night.

From here we could see eastward over the many buildings to the bay where ships were at anchor. The tremendous heat from the fast encroaching flames caused the air to be pulled in from the side, and as the flames consumed the oxygen, they sent its smoke, chimney-like, in a large column upward.

We watched for several hours into a night nearly as light as day. My guess would be that the city blocks in this area were consumed and carried away as ashes, advancing at the rate of a block by the hour, as the flames ran their gamut. It was bright enough for me to take pictures by time exposure at night.

The advancing flame progressed in a huge circle centered on the tall buildings of the downtown section.

After spending this day and a half in San Francisco, we rode on our bikes to where we met a train, and a kindly conductor gave us a ride on the platform of the day coach to Palo Alto, accepting what small change we had left.

Two weeks later we went back with a Stanford group to aid in relief duty for the suffering people in San Francisco. We were under the direction of the Red Cross and were assigned individually to patrol given areas of so many blocks. As said before, no lights and no cooking were allowed in any house or restaurant day or night for fear of fire. I was elected to patrol nearby neighborhood residences where four-legged wood stoves were at the adjacent curb and cooks were busy sizzling steaks and food. I happened to be in an area where people were self-organized with plenty of help and where, luckily for them, their homes were not destroyed. Most of our group of relief workers had similar reports from their respective territories.

Food distribution centers were quickly established and people in want had all the necessities of life with little effort on their part. Things just seemed to flow properly in a well organized pattern, thanks to the Red Cross and the Presidio soldiers in charge.

At night the city was in complete darkness. The burned out area below us was a no man's land. Now and then we could hear a shot. There was a lot of looting going on, and no doubt these shots were fired to discourage people from the temptation of treasure hunting in the burned area at any time.

Ed. Note - For a brief explanation regarding Breer's career between 1909-1916, refer to his biography on page 1.

As late as two weeks after the quake, cooking within the homes or buildings was not allowed for fear of further fire. In this photo, the chef of a popular restaurant was forced to prepare his food outdoors while waiters carried the hot portions inside to the customers. (Breer Collection)

Soldiers were called in to protect incoming relief supplies and prevent the spread of looting, especially since interior lighting was not permitted at night to prevent further incidents of fire. (Breer Collection)

Quake damage extended to the Stanford University grounds in Encina where the huge Memorial Arch lost much of its stonework. (Breer Collection)

The Stanford library building was virtually demolished by the quake. Breer called it a testimony of not allowing a tall mass to move in its natural frequency without frictional resistance. (Breer Collection)

One humorous incident was the head first dive of a statue of the famed naturalist and geologist L. Aggasiz who had toppled off of his second story perch from the quake motion. (Breer Collection)

Part II: With Zeder and Skelton at Studebaker, 1916-1918

Studebaker Research Engineering in 1916

Being single and ambitious, it did not take me long to accept my friend Fred Zeder's invitation to join him at the then young Studebaker Corporation. Before leaving California, I visited the nearest Studebaker showroom in Pasadena. I was impressed by the unusual way in which they displayed their newest model, tipped on its side in the showroom window so that the bottom of the car faced outward. On the showroom window itself, directly in front of it, were fastened large cardboard discs, each calling attention to a special feature of the car to which the disc was attached by a ribbon.

The discs called out the following details: a 354 CID Six-Cylinder Engine, Four Point Suspension, Replaceable Cylinder Block, Novel Splash Lubrication, Pressure Oil Gauge on the Dash, Schebler Air Valve Carburetor, Remy Ignition, Willard Storage Battery, Cushion-Cone Clutch, Worm and Sector Steering Gear, Roller Bearings, Strut Rods from Rear Axle to Frame, Spicer Propeller Shaft, Military Artillery Spoke Wheels, Goodyear Clincher Tires, All-Steel Rear Axle Housing, Bevel Gear Drive, External and Internal Rear Wheel Brakes, Stewart Warner Vacuum Tank Gas Supply, Steel Clad Wood Body, and Genuine Hand-Buffed Leather Seats with Hair Padding.

This list gave me the impression that I was looking at a high grade, dependable vehicle that represented the latest in engineering technology.

What a surprise I received when I later learned firsthand how erroneous this impression really was.

I was put on the Studebaker payroll starting November 15th. I initially stayed as a guest boarder at Fred Zeder's house until I found my own place to live. (Inci-

```
WESTERN UNION TELEGRAM
NEWCOMB CARLTON, PRESIDENT
GEORGE W. E. ATKINS, VICE-PRESIDENT    BELVIDERE BROOKS, VICE-PRESIDENT

RECEIVED AT MAIN OFFICE, 26 SOUTH RAYMOND AVE., PASADENA, CAL
167GS DZ 54 NL
            DETROIT MICH OCT 2-16
CARL BREER
      MARVISTA AVE PASADENA CALIF
WE ARE STARTING RESEARCH DEPARTMENT COVERING ALL PHASES OF AUTOMOBILE
ENGINEERING AND PRODUCTION CAN OFFER YOU CHARGE OF THIS DEPARTMENT
BEING NEW WILL HAVE TO START AT LOWEST COST POSSIBLE
WHAT WILL YOU BE WILLING TO START FOR EXCEPTIONAL OPPORTUNITY
TO DEVELOP INTO BIG POSITION ANSWER IMMEDIATELY YOU CANT AFFORD
TO OVERLOOK THIS OPPORTUNITY
                          F ZEDER
                               1001P
```

Copy of telegram to Carl Breer from Fred Zeder requesting his presence at Studebaker to establish its Engineering Research Department. Breer's acceptance was key to forming the Zeder-Skelton-Breer engineering relationship that became so important in the founding of Chrysler Corporation. (Breer Collection)

dentally, the ground floor of the house where I roomed is now a part of the foundation of the General Motors Building.) The executive offices for Studebaker at the time were on the fourth floor of their large building on Piquette between John R and Brush. Engineering was located here as well as some parts of manufacturing. The major manufacturing and assembly operations were located adjacent to the Detroit River and railway facilities in the southwest part of Detroit.

At that time, the parent company, the famous Studebaker Wagon and Carriage Works in South Bend, Indiana, continued mainly to build wagons and carriages. It contributed very little to the Detroit Automobile Division.

Studebaker production was somewhere between a hundred and two hundred cars a day. This was quite a high number for there not being as yet a moving assembly line at Studebaker, especially when it took several weeks to do a good paint and varnish job on each of the car bodies, which called for extensive floor space.

This mammoth eastern production industry was something entirely new and strange to me at the time. Fred's instructions were simple, "Take ten days or two weeks and look around through all the manufacturing operations before you take hold of the laboratory operation." As a matter of fact, I think this was the one and only directive I ever had from Fred during our entire career. Somehow it just seemed that we had a sort of common understanding, a sort of 60-90 spirit of doing more for your fellow man than one expected in return.

When I first arrived, I glanced through the silent static laboratory. Then I went to the other end of the building where I met up with the operating force of engineers. As I remember, there were five, two of them leaning back in their chairs with their feet on their desks. I explained my mission to them and asked about the nature of the tests they were running. Max Wise, the engineer in charge, replied that they had conducted a number of tests in the past, sent them up to the fourth floor engineering office, and never heard a word in return. His attitude told me everything: either he had a lack of interest in his work or a lack of follow-up. Probably both.

Engineering headquarters was on the fourth floor adjoining the executive offices but the mechanical lab was in a small single story building.

I have never forgotten the first day I took the elevator down to the ground floor, crossed the yard, and went into the engine dynamometer room (the so-called mechanical lab). My guess was that this room was less than twenty by forty feet square. At one end was a "Diehl" dynamometer. Next to it was a "Sprague" dynamometer. They were set up so that their rotating shafts were in tandem with each other.

I suppose I should have expected such a minuscule lab effort. Laboratory activity at the time was not considered important, productive, or even necessary. In those days it was not as important that vehicles go fast as it was that they somehow be kept running to reach their destination. Endurance at high speed was not a requisite.

An ambitious driver of that era would find it too rough and tiring to drive fast. If he persisted in doing so, his engine would soon begin to balk, forcing him to slow down to comply with its natural, horse-like carriage gait.

The First task: Fixing the Studebaker Mainline Engine

After a few days, I began looking into the Studebaker engine currently being produced. It showed 26 hp for a 354 in^3 displacement — a relatively large engine of 3 7/8-in bore with a 5-in stroke.

The engine was hard starting and failed to fire its six cylinders uniformly. The updraft carburetor, especially when cold, sent its major share of gasoline straight up to the center two cylinders via the twin inlet port. As a result, the center two cylinders began to fire first. By the time the twin-ported cylinders at each end received their firing mixtures, the center cylinders became overly rich, fouling their spark plugs.

There were two definite changes to make. The first was to place an obstruction in the path above the carburetor so that the heavy gasoline ends would physically split in the direction of flow. This would distribute fuel more uniformly to the three inlet ports, each of which fed two cylinders. Second was to make this obstruction hollow and direct exhaust gas heat to its so-called hot spot by means of a controlling valve.

The first experimental change our laboratory made gave us great encouragement. The engine's warm-up period was greatly reduced, and the engine began to fire more uniformly.

Next came the task of increasing the engine's horsepower. After various modifications, we soon had it producing 65 hp, quite a boost from the former 26. However, that brought another concern — whether the engine structure could take this increased torque and horsepower load over time. Its structure seemed weak, appearing to weave and bulge because of the sudden added push from each cylinder.

After much experimentation, we added wide ribbed flanges on each side of the crankcase near the bottom that tapered wider from the front toward the wide clutch housing at the rear. Along the outer edge we built in a vertical rib for increased rigidity. The cross ribs to the crankcase and the vertical tie webs in the valve chamber made a tremendous reduction in vibration.

Although we were preoccupied with interesting engine exploratory work, we were asked to evaluate what was known at the time as the "Radcliff transmission" whose one-year option with us was about to expire. A decision on its merits had to be made before the annual New York Auto Show in January 1917.

We immediately contacted Mr. Radcliff and his engineer, and we learned that his transmission had encountered numerous difficulties during road tests. The unit would heat up, and smelly oil vapors would permeate the atmosphere. Also there was no noticeable increase in performance as claimed.

Radcliff and his engineer were very enthusiastic about their transmission in which a hydro impeller replaced the cone clutch that multiplied engine torque. After examining the drawings, I was bothered by Radcliff's insistent claim that his transmission increased torque output similar to that of a regular gearshift transmission.

In essence, the Radcliff transmission consisted of a cast aluminum housing with radial vanes. This housing rotated as a part of the flywheel. Inside and opposite it was a rotating set of vanes connected to the propeller shaft driving the car. When bolted together, the two-piece housing was filled with a special light oil which in turn delivered engine torque to the propeller shaft.

We set up the transmission between our two already in-line dynamometers, using one to measure torque input and the other to measure the torque output.

We soon found that the transmission losses were rather severe. As fluid temperature built up, the output shaft would slow down quite rapidly in proportion. Finally, we decided to make a wide open power run on a Saturday morning inasmuch as Mr. Radcliff wanted to witness these tests but had to take the train that night for New York, and hoped to be at the Statler Hotel by noon. To our surprise, the output torque on the absorption dynamometer became greater as we increased the input load. What's more, the output torque was twice that of the input torque. Mr. Radcliff's enthusiastic, parting shout was, "Just as I told you, we always knew there was an increase in driving torque." We were perplexed, and could not understand how the power run would result in an increase in torque contrary to our earlier tests.

During lunch we plotted the input-output curves. The graphic picture rather quickly revealed what had happened: The culprit turned out to be three innocent spring leaves extending radially from the dynamometer, the ends of which rode between two adjustable stop screws. These were safeguards to cushion sudden overload shock in either direction away from the Fairbanks-Morse beam scales. We found that if we adjusted the stop screws so that the scales would register the added torque that the stop screws were stealing away, a different picture emerged. We called Mr. Radcliff at the Statler to tell him the facts before he got away to New York, but he preferred to retain his earlier impression and cast our findings and facts aside.

The end result was that the Radcliff transmission was discarded, and nothing ever became of it. In essence, it was a very inefficient, mechanical expression of the fluid coupling and was not recognized to be of commercial value until after metal stamping and welding techniques had developed some 20 years later. The evolution of this development led to torque converters with various added runners leaning on in-between gearing to noticeably multiply torque output, thus leading to the automatic transmissions of today.

Searching for the Cause of High Oil Consumption

In those early days oil consumption was great, seldom over 500 miles to the gallon, and more often 200 or 300 miles to the gallon. Piston rings were rectangular with sharp edges. Each piston was made of cast iron and equipped with three rings. A fourth ring would mean a costly design change, and would add undesirable piston weight. We began to experiment with the rings, angling the corners and grooving the faces to increase the surface pressure. We began to realize that the oil splashed on the cylinder walls worked its way up into the combustion chamber primarily around the back of the piston ring. Much less oil was slipping or creeping up along the cylinder wall. By notching a rectangular groove about 3/64 x 3/64 in around the lower outside face of the ring, something new occurred: The notch would scrape the surplus oil from the cylinder wall on the down stroke, and pocket it in the groove instead of pushing the oil back into the space behind the ring. At the bottom of the stroke, most of the oil would again be deposited on the cylinder wall. We soon found that two rings were better than one, and with three rings, the engine would run sufficiently dry to score the upper end or the cylinder bore. If we turned the three rings over, grooves up, the engine would pump oil into the combustion chamber faster than it could be consumed. Then we tried various types of angular slots at the lower outer ring edge. It was one of these that became adopted for production. With it we could get 1,000 or more miles per gallon of oil.

Curing the Self-destruction of Main Bearings

Another problem our team had to conquer was the burning out of engine main bearings in the field because of babbitt melting. At that time crankshaft bearings were not machined to close tolerances like they are today. The practice was to clamp babbitted bearings tight on the engine shaft without oil and by means of a 15 hp electric motor revolving the shaft at relatively low speed. The friction heat would melt or soften the babbitt sufficiently to leave a smooth, finished surface.

We learned by experience that a flood of oil poured on at the right time would cool and lubricate the bearings for a further run-in and burnishing period. This cut down materially the amount of work needed to hand scrape bearings for final assembly.

The important phase of this process was that the tin antimony ratio, and more importantly the elimination of lead, had to be held very close. The specifications called for there to be no "trace of lead."

In our first investigation of a field bearing failure, we found more than a trace of lead. We rushed over to the General Aluminum and Brass Company, the makers of the bearings, and discovered their checking laboratory to be dark and dirty and enshrouded with spider webs. Their chemist had got into the bad habit of writing "trace" behind the word "lead" on the quality check analyses sheet. Later we found that this trace was not the seat of our trouble, but the experience did force the supplier to recognize the importance of proper laboratory technique which proved of great value to their future.

One day we had a brilliant idea: Why not put some windows in the sides of the crankcase and electric lights inside to see what was going on? I will never forgot the surprise we received after we had set up a crankcase accordingly and started to motor the engine on the dynamometer stand. We started the engine slowly for fear if revved too fast, oil would splash on the windows and we could not see. To our surprise, no oil splashed until we were up to 600 rpm, and only then did an oil fog appear. It became obvious that the main bearings were not getting so-called "splash" lubrication, but only "fog" lubrication. We wondered how the engine got along at all. Here were the main bearings continually looking for oil, but receiving only oil fog.

It might be well to describe the Studebaker oiling system at this point. It had a twin gear oil pump that drew its oil from the engine pan-sump and delivered it to a half-inch brass delivery tube that ran the length of the crankcase. There was a lengthy restricted orifice at the entrance end of the tube to indicate oil circulation via an expensive, dash pressure gage. This, as I remember, was costly — something like $2.50 each. There was an inner pan just below the connecting rod ends with oil troughs from which the revolving rod ends could pick up oil to splash the main bearings. The oil troughs were individual with overflow openings such that each rod was supposed to haul the same amount up or down hill. The connecting rods were ingenious in that the projecting dipper ends consisted of two spoonlike scoops that projected from a large central hole in the cap of the rods. These stamped twin scoops curved outward in opposite directions toward the dipping ends. The design was to ensure foolproof scooping should the rod be assembled either one of two ways. The rod bearings received lubrication through the cap openings from which the scooper stampings projected.

The engine's four crankshaft main bearings were cast at the bottom with four cross webs making a box structure with the two long side walls forming the crankcase. The crankcase had a separate cast iron casting to which the unit six cylinder block was bolted.

Along the top edge of the webs two flanges forming a v-shaped trough acted both as a stiffener and a catch-all for the supposedly splashed oil which could enter an angular drilled hole for purposes of oiling the four adjacent camshaft bearings.

Between the v-shaped trough and the main bearing and on each side of the vertical cross webs were two vertical main bearing stiffening ribs, with a dam at the bottom forming pockets for oil to splash in and feed oil to the main bearing surfaces through vertical drilled holes.

What we saw through our window into the lit crankcase were dry oil pockets and bearings. Of course, the upper troughs also were high and dry.

What we discovered contrary to expectations was that oil, unlike water, would cohere and not splash as the rod dippers sped through it, leaving only what appeared to be a groove. The dippers repeatedly coming by appeared to push the oil aside. Of course, enough oil would adhere to the dipper to creep up and satisfy the connecting rod surface but the main and camshaft bearings would starve, burn out, or score. Our answer was to drill fair-sized holes straight through the center of the four camshaft main bearings. These formed oil passages at right angles through the center line of the camshaft from face to face. A smaller hole was drilled downward straight through the respective crankcase four camshaft bearings to the oil supply line located diagonally just below. This hole would feed a spurt of oil through the camshaft into the before-described trough above twice per camshaft revolution. At the center of the trough immediately above each main bearing we drilled a larger vertical hole, sufficient in size to leave an opening on each side of the vertical central main bearing web below. These holes were so located that the oil in the trough would run down both sides between two adjacent webs directly into the engine main bearing oil pockets already there. It was a happy feeling to see how well this worked. The camshaft was constantly lubricated and the main bearing oil pockets were always overflowing with oil. It was also interesting to note how this overflow of oil would follow out on the revolving crank throws, spray off and lubricate the valve stems, tappets, and cams.

Fortunately for us (and Studebaker) not a single change in design or parts had to be made except to add the drilling of a relatively few oil flow holes. Never again did this Studebaker engine starve for oil. What was surprising to us later was that one of the largest producers of low priced cars continued to use a splash lubrication.

Eliminating Valve Train Noise

About this time we found that our valve tappets were getting noisier. We had always accepted a certain amount of "rat-a-tat-tat," but now valve clatter was becoming worse and more annoying.

We found that our valves were manufactured on a piece work basis, and that someone had the brilliant idea of speeding up the heavy "Landis" cam grinders to make more money without increasing cost. These grinders had a very heavy oscillating beam which carried the rotating camshaft back and forth against a large grinding wheel. Oscillation was controlled by large machine master cams which determined the shape of each individual engine valve cam. In speeding up the heavy reciprocating beam, the control arms could be sprung to the extent that the ground cams took on a distorted shape. We also found that the skilled craftsmen would round the cam noses in their polishing and finishing operation because the inertia overswing of the Landis grinders would leave a pointed cam nose. The shape of the cam portion that determines the speed with which the tappet slows the valve when it seats also was of no consideration. After much work with indicator dials we proved to the shop where they were wrong, and the necessity to making the cam shapes according to the drawings.

Although this change meant less valve clatter, some noise was still there. Further investigation indicated that the noise came from the poppet valve slamming down on its seat at too high a velocity, then coming up too rapidly to lift the valve. The answer was simple: Bring the valves down to a stop about six to ten thousandths of an inch above their seats, pull the tappets out from under, and let the valve springs force the valves already in a stop position to their seats below. We experimented by taking a standard production camshaft, grinding the cam bottom diameter smaller between the valve "opening dead stop" and the "closing dead stop" position, giving the tappets ten thousandths clearance when the valves were closed. The valve train noise disappeared miraculously. The valves would come down to a dead stop; then the tappets would step out from under leaving the ten thousandths between. Then at the proper time, the tappet would come up under the valve, and carefully and quietly lift it off its seat. What a tremendous difference the change made. All cams now worked in harmonious quietness.

Ultimately our research forced the development of better production machines and paved the way to the very fine precision machines that now hold cam contours closer to $\frac{1}{10,000}$ accuracy rather than the $\frac{1}{64}$ of an inch common in those early days.

Curing Crankshaft Harshness Noise

Crankshaft harshness noise appeared at several definite engine speeds. It would come out as a sharp peak, then disappear as engine speed went up, then come again at a higher speed, and so on up to even a third top speed. The higher the speed, the louder and harsher it became.

To the non-technical mind, I might explain that the steel out of which a crankshaft is made has elastic properties similar to that of rubber. A seemingly rigid crankshaft will spring in a torsional manner under the pressure of the engine explosive forces.

Every crankshaft with a flywheel fixed solid to it has a fixed natural torsional vibration frequency of so many swings per second. At various engine speeds where these explosion forces synchronize or tune in with the natural frequency, the angular twist of the shaft is amplified, like pushing a boy on a swing. One push every time he comes back will make him swing high, but a push every other time will decrease swing height, and a push every third time will result in even less height.

The same action takes place with the crankshaft. When piston force pushes behind each torsional vibration movement, the crankshaft will vibrate at maximum amplitude, called the primary frequency (a condition that broke many a crankshaft in the early days because of fatigue). The engine speed at which piston push happens every other time is called a secondary frequency. A third one occurs at still lower engine revolutions. In between these are other and less important frequencies.

Returning to the boy-in-the-swing analogy, should we put a pool of water under the swing and attach a paddle to the seat so that this paddle would drag through the water each time the boy went forward and back, we soon would find, regardless of how hard we pushed, that he would not swing at all.

And so it would be with a crankshaft vibration dampener.

To attack the problem we examined an English development known as the "Lancaster Dampener," a device attached to the crankshaft at the opposite end from the flywheel. It consisted of a pair of relatively small disc flywheels that were pressed apart by springs between, and free to revolve except for the friction against two fixed crankshaft retaining flanges. Hard molded composition furnished the desired vibration-retarding friction.

We designed and applied various samples and found them very effective in eliminating the annoying resonance as well as the possible damage from crankshaft failures. Then we took Max Wollering, the production head, and various members out for demonstration runs. Everyone except Wollering felt there was a marked improvement. Next we took Sales Manager Biggs and his assistants out for demonstrations, and they too felt there was marked improvement. Max stood alone, but was in the dominant position of calling the turn.

We then set an engine on the dynamometer, marked the flange and dampener flywheels with radial chalk marks, and ran the engine slowly through its critical speeds. With the aid of the instantaneous light of the electric spark in a darkened room the running engine had the appearance of standing still. We demonstrated how the dampened flywheel would slowly revolve in a clockwise direction when just above the critical speed, then counterclockwise when just below in speed. We had Max observe what we had to offer, and again demonstrated the noise with and without the dampener. Max still was not sold.

After several months, the majority in the sales department and others put on the pressure until finally the decision was made to release the dampeners for production. We later designed more complicated dampeners of greater effectiveness until finally our engineers developed a very cheap and efficient design which involved only a single flywheel bonded by synthetic rubber to a flanged hub fastened to the crankshaft. The bonding medium through sufficient internal hysteresis, so-called internal friction, did a remarkable dampening job. This type of dampener is practically standard on most motor cars of today.

Research Developments Regarding Universal Joints

Another problem we tackled was to find out what was wrong with our "Spicer" propeller shaft universal joints. They were in constant trouble, and wore out too quickly as reported through service complaints. This research led to the start of our mechanical laboratory division.

From my experience at both Stanford and Allis-Chalmers, we conceived the idea of building a very simple machine for testing propeller shafts by running two shafts parallel, geared to each other at their respective ends. In place of circulating current we simply circulated foot-pounds of torque at any desired speed.

We built our first test machine at negligible cost using a discarded machine lathe. One gear box was attached at the lathe headstock end, and had a scale beam arm projecting outward on which a hanging weight would designate the foot-pounds torque circulating through both propeller shafts. The other end gear box was fixed to the lathe cross carriage in place of the tool post. Thus with this simple setup, we could operate universal joints at any angle, speed, or load. All that was necessary was to drive the shafts with a variable speed electric motor. The rotating power required was small, just enough to take care of the rotating friction.

The equipment proved very satisfactory from the start, and later was refined by adding various jouncing motions to reproduce and exaggerate the propeller shaft action when being driven under the car.

The propeller shaft universal joint test setup devised by Breer out of a discarded machine lathe. The machine led to the development of the then new Flick universal shaft design, the supplier of which became the Universal Products Company, later absorbed by Chrysler Corporation. (Breer Collection)

The "Spicer," a popular make of universal joint, consisted of a forged, pivoting cross with machined bearing ends concentric and at right angles. The cross was centrally mounted in two sets of bushings, one set in the driving end and the other in the driven shaft end, so arranged to concentrically keep the driving shaft rotating in center line with that of the driven shaft. This housing was spherically shaped at its outer end. Fastened inside as a unit of the driven shaft was a spherical-shaped stamping which allowed for the angular movement, kept the dirt out, and sealed the lubrication grease in, having the help of a felt seal between the two adjacent spherical surfaces. Not far from the flange end a unique 1/8-in pipe plug was provided with an ample screwdriver slot. When screwed into place the raised stamped flange with side holes provided means for a cotter pin to lock the slotted filler plug into place so that there was no chance of its coming out in operation. The manufacturers considered this a very satisfactory means of sealing in the special lubricating grease.

What caused most of the trouble in the Spicer shaft was this simple grease filler plug. For lubrication Spicer insisted on their specially developed brand of a fibrous grease.

We soon noticed that the universal joint ends would throw or spatter grease at us when we rotated them in our machine. In checking the tightness of the flanges, we could see how grease could get out of the overhanging spherical enclosure. It was the pipe plugs. Although secure enough to hold themselves in, they let the grease out because the threads were a sloppy fit and shaped poorly. Due to centrifugal force, the warmed up grease needed only a very small opening to be driven out.

Fortunately our setup made it possible to develop with confidence a new and novel universal shaft design that was brought to us by a Mr. Flick who had a small design shop. Flick's shaft cost less, was lighter in weight, and would allow the shaft to move endwise in its housing under full load with minimum friction via a ball mounting trunnion. This change eliminated the annoying, clunky noise that occurred when we were bouncing over the highway. (The Spicer shaft had a multiple shaft spline slip joint to allow for end movement. When under load, the high static function would let go to the lower friction in motion and cause a lurch, hence the clunky noise.)

This new universal joint was really developed for commercial purposes on our test equipment. It resulted in the build up of an important supplier company —the Universal Products Company, later known as the Detroit Universal Products Corporation, owned by the Chrysler Corporation. Flick's ball joint design with its freedom from spline friction continues to be popular today, and has no substitutes to date.

Early Carburetor Research and Development

Carburetors, in those early days, were worked out and adapted to car requirements by trial and error. The then known carburetors were Schebler, Rayfield, Johnson, Zenith, and a newcomer — the Ball and Ball. Detail instruction books went out with each car to tell the owner how to make adjustments.

At Studebaker we were continually asked to test many of these devices, all of which invariably claimed to double the number of miles to the gallon. We usually had two or three cars and work stalls where the inventor could install and tune up his device for test. When ready, a test run was made with a fixed quantity of, say, two quarts of gasoline. Every carburetor invariably would provide about the same miles per gallon that we received with production equipment. Eventually we developed a simpler, more effective method of testing and showing comparative results. Before too long the pressure of testing new carburetors tapered off.

About this time, in the state of New Jersey, a father and son, Messrs. F. H. Ball and F. A. Ball, who had a successful business manufacturing the famous high speed "Ball Steam Engine," decided that their type of high speed steam engine was on its way out. They became interested in developing unique carburetor devices, and approached the problem on a scientific basis rather than the common hit-and-miss trial method too common in that day.

When we discovered them, we found that they had a most interesting laboratory at the Penberty Steam Injector Plant on Holden Avenue in Detroit not far from our plant. They were developing their "Ball and Ball" plain tube, two-stage carburetor with fixed jets. In principle, the engine throttle would open gradually, utilizing the full capacity of the first stage, then by further throttle movement, would engage a second throttle adding the full flow capacity of a larger second venturi coming in at an angle between the primary venturi and the main upper throttle blade. The Ball and Ball design was unique. Most carburetors had adjustable needle valves to control the gasoline flow. The Ball and Ball introduced a fixed size orifice that could be changed as easily as removing one cap screw and replacing it with another. The given size of gasoline flow orifice was self-contained as a part of the cap screw-like part.

What intrigued me was their ingenious pioneering laboratory setup. They studied their problem from a basic chemical standpoint, namely, that gasoline consists of various hydrocarbon constituents which, to burn efficiently, must combine with a definite proportion of the oxygen of our atmosphere. The Balls' laboratory set up enabled them to pull air through the carburetor as if they were operating an engine in the car. The volume of air pulled through and the volume of gasoline drawn in for a fixed interval could be accurately measured. By taking a number of gasoline/air ratio points they plotted a system of curves which graphically illustrated the limitations of the engine requirements. Such work by the Balls eliminated all the mystery and wasteful hit-and-miss exploratory work of traditional carburetor designers.

The test apparatus was simple and ingenious. A carburetor to be tested was fastened at chest height to a flange the same as on a car. An inverted U-shaped pipe from this flange extended downward through a circular metal base with a channel around its perimeter that was filled with water. The water acted as a seal allowing only air from the adjacent measuring displacement tank to enter the carburetor.

The air measuring tank was hung from a center rope that was slightly over- counterbalanced by means of a rope, pulleys, and counterweights. The lower, open end of the air displacing tank, when in the up position, was slightly submerged in the larger tank full of water. A large connecting tube, extending above the water level in the center of the adjacent tank, furnished the passage for the air to flow freely to the carburetor chamber. Steam jets produced the vacuum to draw the air through the carburetor duplicating the car engine in action.

To plot the gasoline and air mixture ratios of any carburetor was relatively simple and fast. With the carburetor throttle fixed in a given open position, and steam jets working, the hood was pulled down to seal off the outside air. Then, by means of stop watches, a given cubic content of air flow was timed. Also timed was the given volume of weight of gasoline, as measured mostly by a volume in a graduated glass tube.

The test allowed two basic curves to be plotted: one showed the mixture ratio for maximum miles per gallon at constant driving speed on level road (the lower curve); the other, or upper, showed the mixture ratio giving maximum torque or horsepower for wide open throttle for any given engine speed as plotted on the cubic foot scale.

These two Balls, father and son, as far as I know, were the first to explore carburetion work on a mixture ratio basis, at least for commercial purposes.

The work that Messrs. Ball and Ball were doing was so interesting and fascinating that I spent the most of two weeks at their laboratory working with them. With glass sleeve spacers between the carburetor and the flange above we could watch the gasoline vapors or mixtures flowing upward with the air stream rising through the carburetor venturi. All carburetors were of the updraft type at that time. The Ball and Ball venturi design was unique in that upward angular tubes projected into the venturi at points of maximum velocity. These tubes delivered the gasoline from an annulus around the walls behind the venturi. The gasoline, metered by a fixed jet orifice then mixed with air, rose into the air stream through the tubes in the form of a fine spray. Under constant flowing conditions the curves told a very complete story. It took the mystery out of finding out what was going on under conditions of sudden throttle changes in actual car operation.

Early carburetor development followed two design trends. One, the "air valve" type, consisted of a poppet valve retarded by a light spring which created a vacuum beyond it, drawing gasoline into the air stream as adjusted by a needle valve. The popular "Schebler" carburetor was a typical commercial example. The other, the so-called "venturi" type, probably was known better as the plain tube carburetor. Popular names were Rayfield, Zenith, Holley, and later Stromburg.

Carburetor operation can be expressed very simply. When driving steady at nominal speed, the inside of the manifold is perfectly dry because the high vacuum above the throttle blade would cause gasoline to be completely air absorbed and sustained as a dry mixture. Under this condition the distribution of a uniform firing mixture to all cylinders was no problem.

However, when the throttle suddenly opened and brought the pressure in the manifold up near to that of the outside air, the already chilled dry walls inside the manifold would act like a sponge. The gasoline of the incoming mixture spray, attracted by the power of adhesion, had to first wet the walls before entering the cylinders. This sudden absorption would starve one or two of the various cylinders causing a lull or misfire. To overcome this it was necessary to force a momentary excess of fuel into the air stream so that what was left when reaching all cylinders was of a firing mixture. The air valve carburetor accomplished this by the lag in depressing the air valve when suddenly opening the throttle beyond it.

The popularity of the "Schebler" air valve carburetor faded as marked improvements were made with the plain tube type. Yet contrary to this, Rex Johnson carried his air valve design for many years in the Cadillac because of top power he could produce.

Plain tube carburetors first took care of sudden throttle opening requirements by filling a predetermined volume annulus around the main wide open venturi jet, receiving its refill supply when emptied through the main jet. When idling, the gasoline was metered by a separate jet and fed into an orifice close to the opening edge of the throttle blade. Then when the throttle suddenly opened wide, the power supply fuel was drawn from the main venturi jet below. The first blast pulled the excess out of the annulus which wetted the manifold walls, allowing a firing mixture to the cylinders without starvation. This system worked until we looked for more power, meaning larger venturis and larger manifold walls.

An engine is like a racing human being. The more air both can breathe, the more power they can produce. In engine words, the greater the restriction to the incoming air the less the power, but the easier it is to handle the fuel.

The next important development step in tube carburetor development was interesting. With the more rapid progress of the plain tube the air valve Schebler was replaced as standard equipment by the more dependable Stromberg plain tube. The Schebler could not keep pace and before long went out of business.

The Stromberg was not without its problems. For example when the dynamometer development of a larger venturi carburetor was completed, Messrs. Sheppard and Scott from the Stromberg Carburetor Company would put their carburetor through its paces, then turn it over to me. I would check the handling of the car. I soon found I could trick the car into showing a flat spot. This would occur at a definite driving speed. The next day they said they had eliminated it. I found it again but at a different speed. This game of "hide and seek" went on for a week at which time I called the turns. I asked them to just tell me what caused the flat spot. Reluctantly they admitted it was due to the fact that it took an interval of time to refill the accelerating well around the main jet causing the main jet gasoline feed to the cylinders to starve under certain driving conditions. What happened was that by accelerating the car quickly twice in succession, there was insufficient time to refill the annulus to supply fuel for the second acceleration, resulting in misfire.

I suggested a separate accelerating well that would fill quickly and independently so as not to starve the main jet. Scott said to Sheppard, "How about the 'H' well we were playing with in Chicago?" They explained that this was a separate column well chamber several inches in height. By using the manifold depression when closing the throttle, the well would operate quickly and independently of the main jet and would fill quickly with a volume of fuel in proportion to the manifold vacuum depression.

This was on a Saturday. I charged them with having the carburetor back from Chicago Monday. The change turned out to be just what we were looking for. Later with this so-called "H" well we found that we could pull the car down in high gear to a complete stall, kick the throttle open and the car would accelerate without a miss. This we could demonstrate while running on the idle jet with the main jet needle valve entirely closed, showing how well the "H" well feed could be controlled to supply the engine with the desired amount of fuel to fire the engine uniformly for several revolutions without a miss before stopping.

This was really a great achievement and it stepped the plain tube carburetor way up in advance. Following this, we continually called for larger venturis for more power in competition to the Johnson air valve carburetor. The next big advance was a contribution of the Ball and Ball group, namely, the mechanical throttle operated, pressure gasoline accelerating pump, which, when suddenly opening the throttle, would instantly spray the atomized accelerating charge upward into the manifold. This more positive quicker spray as well as refill action made it possible to greatly enlarge carburetor venturis.

As a result of this accomplishment the last of the air valve carburetors dropped out of competition in the race for more power. The standard carburetors of today are only a detailed refinement of what we established at that time.

This carburetor laboratory work was most fascinating and of course a most important foundation of knowledge from which to build. It just was so interesting that I could not get away from it, so I spent a lot of time with F. A. Ball and his father, F. H., in observation and discussion of the various phases of carburetor operation. F. H., about twice as old as I was and of very keen mind, never forgave himself for accepting an idea I propounded, namely in the sudden opening of the throttle the construction was such that the first burst of gasoline from the venturi spray nozzles was so rapid that it practically moved out ahead of the air stream and therefore was especially effective as an acceleration shot. The next day the very first thing F. H. did when he saw me was to countermand my suggestion—good naturedly, of course. He said he could not forgive himself for accepting my theory the day before. To me this illustrated the sincerity of this logical-thinking man.

Our most pleasant and learned association carried on for many years. When F. H. passed away it was continued with F. A. Ball and grandson, Tom. Later when F. A. passed on, and Penberty Injector stopped building carburetors, Chrysler Corporation took over the eight patents and we put Tom Ball in charge of carburetor development and design.

At Studebaker in turn we soon added a carburetor test laboratory duplicating that of Ball and Ball except that a Nash pump was used instead of a steam jet to create the manifold vacuum simulating that of the engine.

Incidentally, while at Allis-Chalmers, my thoughts were in the direction of developing a better and more efficient way to remove the air from the Tomlinson water spray steam turbine condensers used with the Allis-Chalmers water jet air ejector. Suddenly I had the brilliant idea of rotating a drum with extended radial fins inside an elliptical close-fitted chamber such that water circulating would follow the eccentric path and thereby create pumping action. The water moving in and outward would act like well sealed pistons and by simple porting would result in a very effective, cheap, and continuous acting pump.

The next big advance was from the Ball and Ball group, namely a mechanical, throttle-operated, pressure gasoline accelerating pump which would instantly spray the atomized charge upward into the manifold when the throttle was opened. This more positive, quicker spray, as well as the refill action, made it possible to increase the size of carburetor venturis.

Because of this advance, the last of the air valve carburetors dropped out of competition in the race for more power. Standard carburetors of today are only a refinement of what we established at the time.

In looking back, it is interesting for me to recall some of the strange things that we uncovered in our carburetor experimental work. For example, soon after making laboratory headway, I met up with a fellow by the name of McGloshen whose specialty was truck engineering for the corporation. He lived less than two miles away on West Grand Boulevard and drove to work daily.

He came in one winter morning with a former model, six-cylinder Studebaker and asked if I could find out what was wrong with his car. It was not running satisfactorily. I suggested he take it to our company car service division then headed by T. A. Miller, but McGloshen said he had had it there many times, and they could not fix it. The engine sounded like a lot of rattling bones when running. We took the car to our labs, removed the engine pan, and found what looked like an excessive content of thin lubricating oil that turned out to be mostly the heavy ends of gasoline. This was a surprise considering McGloshen's reputation as our truck engineer. It was typical of what a car owner would experience in the winter with the combination of the Schebler air valve carburetor and the then production cold inlet manifold. Mac did all the rest with the hand choke. He would start the engine, attempt to fire a couple of times, after which it would stall. He would jump out, raise the hood, and give the air valve spring tightening knob a couple of turns, jump back in, and start again. Finally, he would get the engine cylinders firing and would get to the plant in this helter-skelter way.

We were astounded to find that the air valve was practically screwed tight and could hardly move at all. If an engineer could be so foolhardy, what would an average owner do? Carburetor adjustment was everyone's prerogative, and owners always seemed to want to try their own hand at it.

We then had an interesting experience with the Ball and Ball carburetor which brought out the value of the results of so-called foolproofing from tampering. A doctor whose close circle of friends all owned expensive competitive cars liked our new line and bought one of our Big Sixes. He was kidded by his associates for doing so because they all had had problems with our precious Six. One week he drove up with his friends on a summer tour of Michigan's upper peninsula. When they hit an area of tough going on all sand roads, his car would sputter, backfire through the carburetor, and stall under the extra demand of low speed lugging power. There was not a thing he could do about it. He was so embarrassed, and so disgusted from being kidded by his associates, that he turned around and drove back to Detroit. He took the car into service to several of our certified service stations, and they all told him the same story: It's all in the Ball and Ball carburetor, and since it is non-adjustable, we can't do a thing for you. In his disgust the doctor called our Studebaker service manager, and after a heated debate, told him that he was through with Studebaker. The manager called us. I told him to ask the doctor to bring in his car to our laboratory and we would fix it without delay to his entire satisfaction.

That very afternoon we picked up his car and drove over to nearby Belle Isle where we quickly discovered the stalling. To cure it we simply installed a set of originally released spark plugs, and the car worked perfectly. We could not force it to sputter back through the carburetor intake again. What happened was the superintendent at Toledo Champion Spark Plug Plant with a serious view to economize eliminated a simple operation to save cost. The center metal electrode as produced extended from the porcelain of the spark plug one eighth to three-sixteenth inches longer than specified. This added length caused electrode temperature to rise enough to ignite the next incoming charge before the intake valve closed. This pre-firing caused a loud sputter or backfire from the carburetor. The sudden increase of pressure on the upcoming piston would stall the engine, and stop the incoming charge from reaching the following cylinder. We made Champion build the spark plugs as specified. It is interesting to note if it were not for the tamper-proof carburetor, the mixture could have been enriched to smother the backfire and cause a lot of damage due to dilution of lubrication by excess gasoline. These and other incidents caused us to concentrate on the continued development of a so-called foolproof, dependable, tamper-free carburetor design.

Another amusing incident occurred as a result of the safeguard design of the Stromberg carburetor used on the Light Six car. Our foreman of the final car assembly and road check came into my office one day all excited to tell me about what happened to him. He owned a Light Six, and at times would leave the car at his home in the garage. In order to keep anyone unknown to him from using or stealing the car he would remove both the idle and main jet adjustment needle valves from the carburetor. If he should need his car while at the plant, he would send one of the test drivers to his home with the needle valves to install so that they could get the car to run. One day, after sending a driver for his car, it dawned on him that he had not given him the needle valves. About the time he was looking for another driver to send out with the valves in came the driver with his car. The foreman opened the hood and discovered to his astonishment that the engine was running fine without any needle valves in place. That was the first he had discovered that we purposely had been putting controlling orifices at both locations so that an excess in rich burning mixture could only happen through maladjustment.

World War I Days at Studebaker

During World War I, Studebaker along with other manufacturers was required to build track-laying army tanks on which to mount light automatic field guns. The tanks were designed to use the engine and chassis parts of our motor cars then in production as much as possible.

At the same time the government asked the automobile industry to build Liberty aviation engines. The first to be developed and tooled was a larger twelve-cylinder engine. It consisted of a 60-degree twin six with an aluminum crankcase and cylinders of single unit construction with thin wall stamped steel water jackets integrally welded. The connecting rods were the fork and straddle rod type. It was an attractive lightweight design producing some 600 hp at low speed, and was used in a fabric-covered, wood frame, DeHavilland bi-plane of English design but made in the U.S.A.

Four auto manufacturers took on contracts to build these twelve-cylinder engines — Packard, Ford, and Lincoln in Detroit, and Marmon at Indianapolis. Fisher Body in Detroit took on the contract to make a proportionate share of the planes.

Operation headquarters was centered in Henry Ford's multi-story production building, located at the northeast corner of Woodward and East Grand Boulevard. Headquarters occupied the tenth and eleventh floors. Jim Heaslip, top manufacturing manager of Studebaker, resigned in the spring 1917 and took over the responsibility for directing all activities with respect to the manufacture of Liberty engines. As he set up the organization, he asked if I would team with O.E. Hunt to handle engineering with respect to the design follow-up of the engines. At that time, the Studebaker organization would not give me leave.

Later, after manufacturing got under way, and engines were being tested, flown, and delivered in finished planes, Heaslip obtained my release on a dual payroll basis, Studebaker and the government. My responsibility was to observe all 50-hour engine tests to find out the cause of so many destructive engine failures. Engines would suddenly fail without any sign of warning on the 50-hour test stand or when in action on the fighting front. In the air a wrecked engine could be destructive and disastrous. The engine would simply explode into parts, the plane would be wrecked in mid-air, and the pilot and co-fighter lost. In World War I there were no parachutes for the pilots nor did the DeHavilland plane come equipped with brakes on the plane's wheels. I used a crew of five or six men to oversee the tests. We would take turns in covering various plants. Bob Daisley was one of the crew, and in later years headed the Eaton Manufacturing Company in Cleveland.

The 50-hour checking approval tests consisted of ten five-hour runs. Five hours of cruising represented about the time a plane could stay in the air before having to come down for fuel. Between flights the engine was allowed to stand and cool down for a period of half an hour. The cycle was to warm the engine up, then run it wide open three minutes simulating take-offs from the ground. Then power was reduced 20% for equivalent cruising speed for the balance of five test hours.

The engine tests were being held at the Lincoln Plant, at that time headed by Henry Leland with Mr. Martin, production superintendent. The Air Corps desired to take Martin up into the air to demonstrate why they wanted the plane's engine to idle at low speeds. Mr. Leland felt that Martin was too important to allow him in the air, but Martin's two associates did not agree. I took them to the Morrow Flying Field for this purpose. After both had been given 20-minute flights, the pilot said, "I have just ten more minutes to fly, why not take a ride?" Not having a permit I hadn't the least thought of flying, but accepted with enthusiasm and climbed up into the open cockpit of the DeHavilland bi-plane. With the helmet and leather jacket on and a husky safety belt tight around my waist, I sat on the single, button-like cushion without backrest mounted on top of a vertical pedestal support. The pilot said, "You better snap on those two safety straps to your belt." I'm glad I followed the precaution or I probably would not be here to tell the story. We sat tandem fashion, the pilot between the engine and the two-bladed wood propeller rotating up ahead. These World War I planes were made of a wood structure covered with varnished fabric.

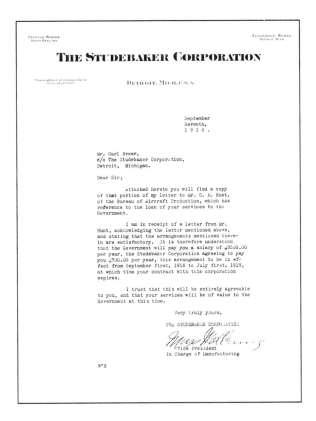

Letter from Studebaker to Carl Breer releasing him for World War I work. Studebaker originally would not let Breer go, but evidently pressure from James Heaslip, manufacturing manager of Studebaker who resigned to head up all activities regarding the building of Liberty engines for the government, brought about the change. (Breer Collection)

The fuselage rested across the lower wing. Above us, supported on wood struts and held by many guy wires, was a second, upper wing. The pilot said, "You may stand up if you want to while we take off." This I did as the straps were long enough to allow standing. As he raised the plane off the ground slightly it was a thrill to be standing there watching the nearby buildings going by while experiencing my first plane ride. Suddenly he stuck the plane's nose upward and I slammed down into a sitting position. Once in the air there was no means of communicating because of the engine roar and the wind that whistled past through the struts and wires. At about three thousand feet he leveled out and waved the plane's tail back and forth and up and down in quick jerks to see if all controls were working. Being satisfied, he raised the nose up and climbed another three thousand feet, then suddenly leveled out and shut off the engine. Engine and air noise quieted down to nothing. Just as I was wondering what happened I saw the earth come at me in the direction of the left wing. The wind started screaming past the struts and whistling at the wires, but there was no engine sound! Something must be wrong! Suddenly, there was a burst of engine noise and I was slammed down hard onto my pedestal seat as the plane leveled out. It was as if the pilot had seen an enemy plane, and had to make a quick maneuver. In reality, he was checking this plane for a stall and tailspin to see how the plane would snap out of it. About that time he looked back and seemed pleased to see that I was still with him.

We then came gradually downward. Circling in for the landing the pilot glided toward the large hanger at the edge of the field. It seemed as if we lost altitude rather quickly, and I noticed that we had just cleared the hangar roof by a few feet. I was not concerned until we were on the ground and Captain Greer, who was in charge of the field, greeted us and remarked that we were lucky. He saw us coming in and drop into an air pocket, and just managed to miss the gable roof atop the hangar. Nevertheless, it was a really thrilling experience for me in the art of early flying...a thrill I am glad to have well behind me. When I went home and told my Barbara of the day's unplanned outing, we both went through the flying thrill again. Sometime later I understand this well-experienced pilot grazed the top of a hanger in landing and both passenger and pilot were killed.

Too many engines just seemed to explode, whether in the air or while being tested for endurance and power at the various manufacturing plants. My instructions were that when an engine failed, it was not to be touched, just left on the spot just as it happened.

One afternoon we got a call from the Ford Highland Park plant where Liberty engines were built and tested. This time the tester had heard a sudden unusual sound, and luckily shut the engine down at the instant it was about to explode.

Upon examination, we discovered the problem to be a localized load fatigue from concentrated, repeated forces on the connecting rod structure. Little was known about the science of metal fatigue at the time. Fatigue was not really evaluated until World War II. What happened was that the lightweight, straddle connecting rod applied its piston thrust against the outside of a split bronze sleeve faced on the inside with a relatively thick lining of babbitt. Due to the flexible lightweight structure of the rod end, the tremendous piston pressure was concentrated over a small area of the bronze bearing shell which was held from turning by the forked rod end rigidly clamped at each side. As the crankshaft revolved, the straddle rod would oscillate through a small angle while applying the piston force. The relatively thick babbitt would get tired of taking the localized load under this continued, concentrated pounding, and a small oblong disc area of babbitt would sheer loose around its edge. The crushing sheer took a definite angle outward of about 45 degrees around the edge and this disc would come loose from its bond to the bronze back. Then the rotating friction of this flake wedging against the crank bearing surface increased the pressure along the trailing edge where, through added friction and lack of heat radiation, it would cause the babbitt metal to flow. Once flowing, it would continue in a narrow path around the bearing surface until it would shut off the nearest oil feeder hole located at the split of the two bronze bearing halves, then it would continue around, shutting off the hole at the opposite split. At the same time, it would clog the lubricating oil coming out of the crankpin. Once this happened, all the babbitt melted, and the pounding would break the piston loose. With the rod swinging free, the engine in so many seconds would be battered to pieces. We finally cured the problem not long before Armistice Day, meaning the war was over before a rod redesign or modification could reach production.

After the war, new Liberty motors were on sale for as low as $100. Those who had the will and know-how bought them and made very serviceable, long-life engines out of them. Many were put in flying service or used for motor boating. In fact, the Allison Company, racing car builders in Indianapolis, rebuilt 500 Liberty engines with copper lead bearings for our government. These rendered wonderful mail carrying service. Even Lindbergh flew them. The results of our findings at least proved to be of service to air mail carriers, and possibly to the Allison Company as well.

Post-war Studebaker Engineering

With the war over, I went back to Studebaker. My aim was to design engines that would run 3,000 revolutions per minute at wide open power for 50 hours without failure. But what I discovered at Studebaker was that they had set up a new checkup department over our engineering division. It was known as the M & S division, namely the Materials and Specifications Division, and it had been put into place without the Engineering Division's knowledge. The intent might have been okay, but it did not work out that way. Engineering releases now seemed to take forever before being approved for production. Also, we found that M & S had installed dynamometer equipment, probably to check our results. We did not like the delay, nor the oppression. One day the head of the M & S division, Mr. Carl Herrmann, and I were discussing various types of multi-cylinder engines. He argued that a four-cylinder engine was in better inherent running balance than a six-cylinder, both of which we were building. This broke down our morale, just to think that we had a department over us with far less good judgment than our own.

We Save Studebaker from Receivership

After returning to post-war engineering work, we were called to rescue the vehicle from rear axle failure. The rear axle was a semifull-floating axle designed so that if the axle drive shaft broke, the wheel could not come off, thus avoiding a possible accident. Struts or radius rods were featured so that in case of spring breakage, the rear axle would stay in line. Added to this was a torque arm that extended forward and linked to the frame to keep the axle housing from rotating when engine or brake power was applied. In addition, the transmission was mounted as a part of the rear axle assembly. The design should have proved satisfactory although costly, but after a large number of cars had been placed in the field, an overwhelming number of axle shafts began to break and come in for replacement. We discovered that the axle housing was weak. If there was a sudden brake application or the car was backed into a curb, the axle housing would become permanently bent due to the severe backward thrust of the strut rods. Thereafter it would act as a fatigue machine and break axle shafts as fast as they were replaced.

The rear wheels were mounted on roller bearings which allowed them to rotate rigidly in alignment with the tubular ends of the axle housing. The axle drive shafts had splines on the inner end to engage the differential. On the outer ends were large, cone-shaped forged flanges integral with the shaft.

When the axle housing would bend, it would force the wheel to revolve out of center line causing excessive strain on the axle drive shaft producing fatigue failure. Once the housing was bent, new replacement shafts likewise would break because of fatigue.

Just at that time we were warned by Mr. Erskine through Henry Goldman of Goldman & Saacks, the financial backers, that more money was going out of Studebaker than was coming in. Unless conditions improved, they were going to tighten up on the purse strings.

This set off a wave of excitement throughout the organization. Frankly it all boiled down to the following facts: The product was not satisfactory in the field; a redesign of the complete vehicle must be accomplished quickly; and there was no surplus money for a costly tear-up that would require new tools or fixtures.

Here we were, a group of engineering youngsters in the automotive field, face to face with a serious situation. With this charge for redesigning the entire product, many things ran through our minds. We had already shaped up as an engineering group — Fred Zeder as vice-president in charge, Owen R. Skelton doing a good job directing and handling design, and I, directing research and running the laboratories. The three of us took on the challenge with enthusiasm.

The 1918 Studebaker Redesign

First a new engine. We were confident that we could produce a very satisfactory unit without the tear-up or additional expense of new foundry flasks.

Because of Studebaker's recent rear axle trouble, Skelton already had been making designs and studies to eliminate its present, costly, troublesome rear running gear. His studies pointed toward great simplification and reduced cost. The transmission would be taken off of the rear axle and placed on the engine or amidships. The axle would be a simple, rugged design that eliminated both the strut rods and the torque bar that were used before. Both the reaction drive torque and the thrust of the wheels on the road were to be taken by the semi-elliptic springs bolted fast to a much stronger rear axle housing. This was a relatively new design for its time known as the "Hotchkiss Drive."

However, Harry Biggs, Studebaker's sales manager, a six foot three, 250-pounder, was opposed to this design. He mentioned that the safety of a "full floating" axle had been shouted to the housetops by competition for years as well as that of the simpler, less costly, "semi-full float" axle.

There was a lot of discussion about the hazards, the relative merits, and cost of the proposed new design. From our viewpoint the new axle design, if properly engineered, would be far safer and a lot more dependable than anything on the road before. The secret was to design the axle shaft so that it would only break in the section inside of the axle housing beyond the bearing, but close to the differential end. Thus, in case of axle shaft failure, the wheel could not possibly come off and cause an accident.

The battle with sales was hot and heavy, but after much debate, Fred Zeder won out over Harry Biggs, head of the sales department, on condition that engineering assume full responsibility. Fred agreed and visited Studebaker dealers from coast to coast on an educational and selling campaign. At the dealer meetings Fred carefully explained how the axle was designed with the objectives of safety and endurance, simplicity, and low cost. He then closed with a unique parting shot:

"Have you men ever thought that the front wheels of your automobiles have always been running on an overhanging spindle? Perhaps you never realized that if the spindle broke the hazard of an accident would be far greater!" So the Hotchkiss design was adopted, and still is one of the top designs today. I doubt if anyone has ever seen a rear wheel broken off at the shaft since.

Regarding the transmission, it was of conventional design, having straight spur gears, three speeds and reverse. The entire assembly was located "amidships" on the chassis. The two parallel channel frame members supported the engine forward and extended back to carry the transmission which was solidly fastened with three straddling arms.

We first tried an aluminum spool coupling between the transmission and the clutch. We gave it radial bolting flanges with Thermond flexible discs at each end. These rubberized fabric discs allowed a flexibility in fastening the inner flanges to the clutch and transmission respectively at each end.

Severe road tests on the initial design indicated a problem. We quickly redesigned the so-called "rag joints" for production in place of the unsatisfactory inner and larger outer flanges to which the rag discs were bolted. In their place we used a four-point lug fastening which bolted alternately to the fabric member. This worked very satisfactorily, giving flexibility and long life.

The Sales Division decided to cover field requirements with three cars — two sixes and one four-cylinder. The advertising terms were going to be "nautical"— the Battleship, the Cruiser, etc. Finally we discarded the nautical terms and simply called them the Big Six, the Light Six and the Four. The Big Six large engine retained our former 354 in^3 displacement size. It advertised 65 hp, over twice that of the previous Studebaker. The Light Six had a displacement of 288 in^3, and the Four 192. As to performance, the Big Six was good and the Light Six fair. The Four did not have enough performance worth talking about, so George Meinzinger, who was in charge of experimental, and I decided to improve it. We substituted a larger bore 3-7/8 x 5 cylinder block from the previous year, and with a few changes in cam shapes, manifold, and carburetor, we assembled in secret a new four-cylinder car that George and I took for a run all through a hilly section of the southwest. It showed a marvelous improvement over the smaller Four, running all over those hills like a "scared wild cat." We had a great new engine to offer that involved no tear-up. The cylinder block, patterns, tools, etc. were as standard as before. Much to our disappointment, we simply could not put it across. The sales division insisted that an improved Four would interfere with Light Six sales. (I have always felt that had we sold that revised Four, the Dodge Four manufactured by John and Horace Dodge would never have become popular.)

Breer, Erskine, and Harold Vance pose beside the 1918 Studebaker Four-Cylinder model during its road test. Breer and George Meinzinger were to redesign the Four into a much peppier engine that Sales refused to accept because it might diminish sales of the new Big Six. (Breer Collection)

The Big Six on the road with George Crist at the wheel during the same test trip. (Breer Collection)

One jolt we received was to discover that in the midst of our design program, a complete, all new six-cylinder automobile car had been engineered for Studebaker by A. A. Smith Company in Milwaukee. It was a secret assignment, probably a matter of insurance. A Mr. Ferguson of Detroit was the designer and builder. By the time the vehicle arrived, however, we already had proven the superiority of our design, which was accepted. Ferguson later was to design and manufacture the Divco delivery wagon, a stand-up drive vehicle that became quite popular on milk and delivery routes.

Soon we had been given the go-ahead to finish three sample cars, one each of the Big Six, the Light Six, and the Four. Each was a closed car. Mr. Erskine had the brilliant idea of driving the three cars to Lake Saranac and presenting them to Henry Goldman on his birthday while at the same time having Mr. Fish (Studebaker's son-in-law) present Mr. Goldman with a gold headed cane. It was a tight schedule to make, giving us little time to check test. It meant shipping the cars aboard the Detroit and Cleveland Navigation Company's Side Wheeler for an overnight trip to Buffalo, New York.

Another two days rolling along cross country and we arrived at the gates of Henry Goldman's beautiful lodge. We drove the cars into the garden at the rear of the boat house, a picture spot with its electric launches and various small boats.

It was a beautiful sunny day. Mr. Erskine and Mr. Fish presented the golden cane with congratulations and best wishes to Mr. Goldman. He showed great interest in the three cars, and asked many questions. He said to Fred, "Are these good cars and ready for production?" Fred nodded with enthusiasm and said that there were none better. There were no more questions. He then took Erskine by the arm and asked us all to join him for a noon day dinner.

Getting Mr. Goldman's satisfied approval on his birthday meant a new day for Studebaker. Mr. Erskine, Clemente Studebaker, and several others left us the next morning while the rest of our crew continued our test run northward to get better acquainted with our new automobiles. We traveled over into the Canadian border, made an overnight stand at the famous Hotel Frontenac in Quebec, and left early the next morning homeward bound. The trip took us away only a little over a week.

Before long we began production of the new Studebaker, mainly sedans, some open touring models, a coupe and possibly a roadster to follow. Our body styling and engineering division was located in Henry Ford's original factory building then a part of the Studebaker plant on the corner of Second and Piquette. J. H.

More of the Big Six on road test. The Big Six was a redesign by Breer of the Six offered by Studebaker in 1917. It retained its 354 cubic inch displacement but produced twice the horsepower — 65. (Breer Collection)

The 1918 Studebakers Six and Four on caravan through a 2,000 mile road trip that took them through Pennsylvania. In the absence of proving ground facilities, road trips were the order of the day for prototype testing. (Breer Collection)

This was a routine scene in the farmlands surrounding Detroit as manufacturers took to their roads to test new models. Each car carried four spare tires in reserve so as to minimize down time. Engineers, in fact, looked for the worst roads they could find to maximize the test runs. (Breer Collection)

A good rain on Michigan back roads provided all the rugged testing a Studebaker engineer could desire. The roads of this era were bad enough under ideal conditions, let alone during spring thaws. (Breer Collection)

Bourgon was in charge. Oliver Clark was his understudy and was just coming up as his assistant. Of course all bodies were made of structural wood, metal-shod, painted and varnished as established by the horse-drawn carriages. Our styling was just getting away from the whip-socket era.

Saturdays and Sundays were part of the work week. We shuttled between our headquarters on Piquette and the main plant across town at the end of West Grand Boulevard on the Detroit River.

Everything seemed to be satisfactorily underway to begin shipping cars to the dealers when we learned that two of the cars had caught on fire. They were Big Sixes equipped with Ball and Ball two-stage carburetors.

It did not take long to find out what was happening. The secondary stage of the carburetor pointed downward at an angle into a pocket formed by the stiffening web along the lower stiffening flange of the crankcase. Vertical ribs and webs to the crankcase side wall formed a perfect pocket in which gasoline could accumulate when spilled out of the carburetor secondary stage. When suddenly opening the throttle of a cold engine at low engine speed under certain repeated conditions, it was possible for the gasoline in the engine pocket to be set afire from a backfire flashing through the second stage throttle opening. Fortunately we quickly discovered that by making an upward, spoon-shape baffle just outside the second stage opening, the backfire flame would be chilled, and any excess flame turned upward away from the pocket making it impossible to start a fire.

Working Saturday night and Sunday at Penberty Brass Foundry, sufficient flame baffle castings had been made by Monday morning to install on the first cars being shipped, and all other cars thereafter.

Incidentally, it was at that time that Walter Chrysler was directing Buick, and found that the new Studebakers were becoming a bothersome competitor.

Part III:
We Create Chrysler Corporation

Our First Contacts with Walter P. Chrysler at Willys-Overland

It was 1919, and after our previous year's success at Studebaker, things were not going as smoothly as we would have liked.

About that time we received word from Walter P. Chrysler through an associate in Detroit, one O. G. Brown, a tailor and a popular friend of every top automotive executive who had sold many stylish, tailored business suits to them. The word was that Chrysler wanted our engineering group to join him at Willys.

Walter P. Chrysler had just left Buick and General Motors after a falling out with Durant and had joined the John N. Willys-Overland enterprises as vice president and general manager in January 1920. After several contacts and trips to New York and to the attractive new Willys automobile plant at Elizabeth, New Jersey, Fred Zeder, Skelton, and I decided joining Chrysler was a logical move to make.

To Elizabeth, New Jersey we went, taking with us a group of 28 men, the cream of engineering talent that we had built up at Studebaker.

At Elizabeth, we joined Jay Hall, who was in charge of all Willys' operations, Don S. Devor, who headed production, and Earl Wilson, who was responsible for sales. Devor had been one of our group of apprentice associates at the Allis-Chalmers Company back in '09. It was Devor who had suggested to Walter Chrysler that he contact us to see if we would help them to solve their engineering difficulties at Elizabeth. John N. Willys at the time was famous for the Willys-Overland cars of poppet valve engine design, and also the Willy-Knight which was built around the famous Knight sleeve valve engine patents developed in England.

Through Devor, we learned that an engineering division already had been set up at the Elizabeth plant headed and directed by a fellow named Segardi, a very fine

Italian engineer. Segardi had been associated with Devor at the Oldsmobile plant in Lansing, Michigan where Devor had been Oldsmobile's production manager. Segardi and his group reported directly to Willys-Overland in Toledo, Ohio, and its engineering division chief engineer, Ed Beldon, and his assistant, Goldsmith. Beldon, formerly body engineer for Packard, was engineer in charge of all the Willys-Overland plants. Beldon's fame came from having developed and produced a four-cylinder automobile equipped with a poppet valve engine that competed with Henry Ford's Model T. A lightweight car on a 100-in wheelbase, it was prominently advertised as offering a wonderful ride by virtue of its "Wish Bone" springs. "Wish Bone" meant two cantilever springs angling back from the frame to the front axle, and forward from the rear end of the frame to the rear axle. They added up to a 120-in spring base on a 100-in wheelbase car. The car became popular, and John Willys built over 100,000 that year. (Ed. Note: There is no record of Willys having built a 100-inch wheelbase car in this time frame nor of Willys exceeding 100,000 units.)

Willys, being a fast operator, decided it would be advantageous to have a plant near the Atlantic seaboard, save the freight, and take advantage of the populated eastern market.

At Elizabeth, New Jersey, a beautiful 12 million dollar plant had been erected by a very able Chicago architect and builder, Paul Bettey. This plant was complete with its own self-contained steam power plant and heating system, and was located at the crossroads of major railroad facilities. (Ed. Note: It had been completed in 1916 for Duesenberg Motors Corporation, but had never seen automobile production because of World War I, and the later decision of the Duesenberg brothers to forego the four-cylinder car that New York capitalists, underwriting the venture, had expected of them. Both plant and corporation were acquired by Willys in 1919.)

About eight million dollars had been spent in tools, dies, and fixtures to build a new six-cylinder model patterned after the apparently successful Four. So here we were. There were sample six-cylinder production cars already built, and as Mr. Devor explained, now being tested. Road tests, unfortunately, had demonstrated that the vehicle's springs were breaking. When they were stiffened, the supporting frame end would break. If both were stiffened to a point of acceptable endurance, the ride would revert to that of an old fashioned lumber wagon.

In analyzing the cause and effects of the spring breakage plus ride difficulties, we soon realized the increased wheelbase (over the four-cylinder car) added sufficient weight to overstress the front spring. The cantilever (wishbone springs) were bolted to a frame front U-shaped cross member which also was overstressed and deflected some, assisting the spring action. Stiffening the spring design to bring the steel within a safe fatigue limit resulted in more stress being concentrated on the frame, causing the cross member to break. Making the cross member of heavier gage metal would further add stiffness to the ride. It was a vicious circle! To save the ride meant that either much longer springs would be required or the load on them would need to be materially reduced. The first alternative would mean a complete tear up in design which was not practical; the second an impractical weight reduction of an already tooled up, inherently heavier, six-cylinder engine.

We quickly set up a modern engine laboratory with two of the latest dynamometers available. As we proceeded with the engine investigation we soon realized that there were a lot of other major weaknesses such as keeping sufficient water in the radiator to keep it from boiling over. Laboratory dynamometer tests showed us air was slipping in, causing water loss through the radiator overflow pipe. The water pump housing was cast as an integral part of the cylinder head. The belt-driven centrifugal pump mounted directly in the cylinder head caused a sufficient lowering of pressure in the water jacket through its suction to allow the outside air to seep in through the copper asbestos cylinder head gasket causing a loss of cooling water. In this case, operating conditions are such that air will leak in where water will not leak out.

Practically all systems today have the pump pulling on the already cooled water at the radiator bottom tank, forcing the water through the engine block under higher pressure. Had this been the only engine weakness we might have concentrated on finding a single solution. However, the next problem we ran into was the intake manifold. Very simply, an updraft carburetor at one end of a cast passage led past the side openings to each of the six inlet ports, ending at the sixth cylinder. It was evident to us that the unvaporized gasoline distribution would be uneven, especially so because there was no manifold hot spot or heat provision. Under cold weather starting, the uneven distribution became even worse.

Another problem was that the engine crankshaft was relatively weak since it had only three main bearings, one at each end and one in the middle. The shaft between the bearings had to take the brunt of three cylinders which, in our estimation, did not add up to a smooth engine. Our dynamometer checking tests verified this line of reasoning. It certainly was a discouraging situation since the patterns, dies, tools, and fixtures were practically all set awaiting approval to go into production. There were only two logical paths that we could pursue: One, to make as much improvement as possible in the present design, the other, to start with a clean sheet of paper. After consulting with Walter Chrysler, we received approval to do both. Which created another problem: If we were to conduct the new design in the plant alongside of work we already were doing to modify the car now tooled for production, many questions would be raised which, before long, might break down the morale of the plant organization.

For this reason we moved the part of our operating group concerned with the design of the new engine to the Beechwood Hotel in Summit, New Jersey, about 10 or 12 miles by motor car from the plant in Elizabeth. Here, in a portion of one of the upper floors, we moved in drafting tables and equipment and worked in seclusion. In fact, some of our men already were living there, while others moved in later.

We were leading dual lives, except only the heads of both sides of the Willys family knew about it. As to using parts of the original design, there was not a thing that we could logically use.

The fallacy of obtaining a wonderful ride by having a 120-in spring base on a 100-in wheelbase car, as the Toledo Willys Four had advertised, had no fascination as far as we were concerned. To us it was only an illusion dreamed up for advertising. As a matter of fact, this particular design of spring suspension was costly and lent itself to difficulties. It required longer side frames which, for rigidity, would have to be heavier. The only thing that could be said in its favor was that it spread the weight further forward and aft, thus slightly favoring the law of "center of percussion" (which no one recognized at the time), and it was an expensive and extravagant way to go about it.

Our alternative design would be substantially different. It would have an "L" head engine with a seven bearing crankshaft, six cylinders in-line, a shorter stroke with respect to bore, an updraft carburetor with hot-spotted intake manifold that would provide a more uniform distribution to the three cylinders on each side, a Hotchkiss rear axle setup, and semi-elliptic front and rear springs.

This called for an entirely new body and hood design, eliminating the so-called "mother-in-law" seats. We set up a body style division headed by Oliver Clark at the Elizabeth plant. Clay models were made as soon as the blackboard, full-size drawings of an open model, five-passenger car were approved. We then had an independent Elizabeth wood shop make a wooden, full-size model of the body, upholstered with canvas fold-back top and jiffy curtains, and with wood axles equipped with disc wheels designed by the Budd Manufacturing Company. This finished mockup was set up in one of the upper Elizabeth plant floors.

Walter Chrysler was proud of the new design. One day on short notice he phoned for us to stand by because he was coming over from New York and bringing John N. Willys with him. As they arrived from the freight elevator, Walter hurriedly led John directly to the wooden model, and in so many words said proudly, "Here it is!" John took one quick look at it, and said loudly, "Have you had it on the road yet?" The incident amused us since it was so apparent, at least to us, that it was only a wooden facsimile.

We continued to follow through with the design and development of the new vehicle in a running prototype. With our test driver, Tobe Couture, at the wheel, we would demonstrate how smooth and quiet this car was with its 56-in tread riding on the 56-in width streetcar tracks, on the cobble pavement of famous Fruselyhausen Avenue in Newark, New Jersey, or over the hills of Orange Mountains. Walter Chrysler became more and more enthusiastic and proud of our new accomplishment sending over many interested people for ride demonstrations. By this time Tobe Couture had developed a new demonstration technique. First he would take the newcomer for a drive on the long empty fourth floor of our factory building filled with round concrete columns about 20 feet apart. At the far end five or six hundred feet away was an elevator shaft surrounded by these columns around which Tobe would make a sudden turn to the right, swing around the elevator shaft, and back up the same straight stretch, and circle about in similar manner to return. In this way he would demonstrate the car's ability to the newcomer at fairly high speed. After four or five runs like this, he would accelerate to a much higher speed and rush for the turn around the elevator. About this time the newcomer would sense danger and shout out, "Tobe, you can't make it." Tobe, knowing all the time he was not going to take the turn, would continue straight ahead and slam on the brakes! Tobe then would demonstrate the car in heavy street traffic and over the highways and steepest grades of the Orange Mountains. Invariably we won the acclaim of the newcomer.

About this time (1919) Walter Chrysler made a sporting bet with Harold Wills, who had developed his own new car, the Wills St. Claire. The sporting event was to consist of competitive tests on a famous long hill Harold Wills had selected nearby in New Jersey, comparing speed and acceleration. It involved some 20 people and a half dozen cars. It was a great day for Chrysler and our comparatively lower cost car which Harold Wills had to admit had outperformed his.

By this time, it had become a well established fact that there was not much of a chance that we could make a commercial product out of the other six-cylinder Willys engine design. It involved too much of a tear-up to fix it, and moreover was obsolete in comparison to the new car we had just developed.

All this had us building up enthusiasm toward putting the new car into production. Already a large sign of incandescent electric bulbs stood on top of the plant spelling out the name of "Chrysler" in huge letters which was the name planned for the new Willys car. (Ed. note: Automotive Industries reported that a Chrysler Six was expected to go into production during the fall of 1920.) Then came a lull which we naturally assumed to be required for planning the steps to be taken in changing over to a new product. When we had joined the Willys Corporation, it was an established fact that there was plenty of reserve capital in the treasury for all such contingencies.

We Leave Willys-Overland

Then one day, like a thunderbolt out of the sky, came the news that the company did not have enough money to carry on, that the Willys-Overland branch at Toledo had been having difficulties and had depleted the reserves set aside for the Elizabeth plant division. In other words, Willys Corporation was headed for receivership. (Ed. note: By February 1922, the Chrysler Six, the Zeder-Skelton-Breer design, still had not gone into production two years after being announced, and Chrysler resigned from the Willys-Overland Co.) It was a tough spot for us to be in, having assumed the responsibility of coaxing a team of fine engineers away from an already prospering plant at Detroit. Suddenly all our achievements and built-up excitement disappeared. Our immediate problem was to see what we could do to keep together the proven and efficient design engineering team that we had worked so hard to build. There was only one answer — to set ourselves up as a consulting firm.

Zeder, Skelton, and I went to Mr. Chrysler and told him we wanted to set up our own consulting offices. He, in so many words, said, "That's fine. You three go downtown and set yourselves up as consultants." We answered that was not what we had in mind, that we intended to keep our entire team with us. Moreover, we needed all of them to handle the detail work for any and all consulting work that we may take on. Mr. Chrysler finally realized our predicament, and admitted that it would be the logical thing to do. In the meantime, Mr. Chrysler had assumed the responsibility of chairman of the board of the Maxwell Company. Therefore, we presumed he would consider us for any design contract for the Maxwell group. If Chrysler could arrange with Maxwell to make us an advance payment, it would help us get started as consultants. Encouraged by the new change of events, we looked for a place in Newark, New Jersey, large enough to move in drafting tables and equipment.

Newark was the logical place for us. It was close enough so that none of the men had to change his residence, was central with respect to railroads, and had the only underground tube transportation to central New York City.

We found just what we wanted — the second floor of a two-story building at 112 Mechanic Street. We moved in the same day that the previous occupant (a furrier) moved out, and immediately established a charter under the name of The Zeder, Skelton, and Breer Engineering Company. To be successful we had to find enough contracts to keep our group busy while quickly collecting sufficient money to meet our heavy payroll. This the three of us planned to do and without reducing the wage scale. We also made arrangements with the Willys receivers to use the laboratories we had already established at Elizabeth for nominal rent.

Since the Elizabeth plant was especially designed and equipped for motor car manufacture, the receivers decided it would be best to auction the entire plant, its equipment, and material at a sale that would take place at the plant. This was quite an exciting day for us, to see who would become the owner of a twelve million dollar plant with some eight million dollars in material and manufacturing equipment, especially since Mr. Chrysler would be involved in the bidding.

Several hundred were present for the auction. At first five or six groups did the bidding which took place in an incognito manner. No one seemed to know whom they were bidding against. None of the top men from the large companies was visible. Surprisingly the entire deal was closed within an hour. The bidding started in the one or two million dollar range, then settled down on bidding increments of 25 or 50 or 100 thousand. Above three million, several bidders dropped out. Finally it was settled without much fanfare at $5,520,000. The buyer was the famous William C. Durant. He sat down and wrote out a single check. At the time it was reputed to be one of the largest single purchase checks ever written.

We were disappointed as a group for we had hoped that Mr. Chrysler with Maxwell-Brady financial backing would get the plant. In fact, the name "Chrysler" already was on the plant in large letters. Later we learned that Chrysler had stopped bidding below the four million dollar mark. Durant was disappointed because he thought that our designs were included as part of the deal.

After the sale, we realized that we were on our own as a group. Mr. Chrysler had arranged to help our consulting business get started, but now we contacted Mr. Durant as well, and had several meetings with him. First he suggested that we take over and reestablish the Mercer car plant that previously had failed. Zeder,

This rare photo was taken of the sign on top of the Willys Corporation New Jersey plant in Elizabeth, New Jersey in which Walter P. Chrysler fully intended to build the car designed by the Zeder-Skelton-Breer team. Willys, however, ran short of funds and was forced to sell the plant for which Durant outbid Chrysler at auction in 1922. Durant made it one of the capstones of his new Durant Motors Corporation. (Breer Collection)

Skelton, and I drove to Scranton, New Jersey, took one look at the old brewery buildings where the Mercer car had been built, and we decided that it was not for us.

Finally, we worked out a contract with Durant for a fixed sum to design a new engine for the Locomobile which was meant to be the high priced model of the Durant Motor Corporation. Several hundred of our new engines were delivered in Locomobile cars before the company was forced out of the automobile business. We heard the few engines they did build turned out to be very satisfactory.

Later Durant decided to build an entirely new plant at Flint, Michigan, and introduce what was to be known as the "Flint Car." We took on the contract of designing the engine which would be built by Continental Motors in Detroit. Durant's forward planning system was to call in all suppliers, have a conference, and practically overnight decide what each supplier was to make, tool up for, and produce in numbers for the new planned car. An initial conference provided main objectives toward which each supplier had to design in terms of what he, the supplier, would consider correct and satisfactory for the job, whether it be an engine, axle, frame, spring, wheel, tire, or body. This was how both the Flint and Star cars were designed and put into production.

While these parts producers were tooling to meet the schedule set, Durant's plant and manufacturing men had the Flint car plant erected from ground up and equipped for manufacture, finish paint, and trim.

With respect to the Flint car, Mr. Ross Judson and his engineers from Continental Motors sat in with Durant and determined the engine size. Given the displacement size, we took on a fixed sum contract to design the engine which Continental would build. Our contract was directly with Mr. Durant. We delivered the complete design details for a six-cylinder, in-line engine with a seven main bearing crankshaft, L-head updraft carburetor, etc. Later, after running into financial difficulties, Durant said the Flint was the best car and engine he had ever produced.

Previous to the setting up of the Flint Company, Durant again called in all the suppliers, and said he had decided to build a low priced car to compete with Ford. This would be known as the "Star" with a four-cylinder Continental engine, standard gearshift transmission, Hotchkiss rear axle, etc. in several open and closed body styles.

This latter effort was to get a car into production at the newly bought Willys plant in Elizabeth as soon as possible. Durant had his right hand production man, Mr. Hoenzie, with his son realign the plant for high speed, progressive line assembly. Production of Star cars rolled along on a high output basis for a while until field condition difficulties banked up and resulted in another company failure.

Now and then we would have occasion to visit the Star assembly line. After seeing men drive spring shackle bolts into place with sledge hammers, we wondered how long they were going to last. So many things happened in the field to Star cars without proper follow-up that the motoring public just quit buying them. Let me cite one typical case. The Star's thermo-siphon radiator was held from fore and aft movement merely by a large diameter hose connection through which the hot water flowed. As a result, engine and car vibration caused the thin brass radiator top tank walls to fatigue. Before long, they would crack and break loose. This might easily have been corrected, but nothing was done, and cars continued to be produced as before. In brief, there evidently was lack of a centralized engineering setup, or if there was one, corrective changes in design probably had been overruled. In our association with Durant we had nothing to do with the Star design.

Meanwhile, our Zeder, Skelton, and Breer consulting business, formed overnight without capital, was doing quite well. We were fortunate to hold all the men who came with us from Detroit, and added two former Duesenberg men — a development engineer, Mr. Tobe Couture, and an engine dynamometer man, Frank Schwartzenberger. Later our metallurgist, Tom Wickenden, had an offer from the International Nickel Company in New York. Since our metallurgical requirements were not an important issue, Tom left, the only one of our group who did so. A graduate of Michigan, he accomplished a lot for International Nickel Company in research and product development.

We had to make a good many trips to Durant's office because he was always in arrears in payments according to our contracts. Either Fred and I or Skelt and Fred would make the trip to New York City.

Some of our visits were interesting. For example, his appointments for us usually were in the morning, so we would take the early tube across to New York, find our way up to his office on an upper floor on Broadway not far from Times Square, announce our presence, and be ushered into a large room around the walls of which were miscellaneous chairs where a number of other people were waiting. At a given time, into the center of the waiting room, a barber would arrive and set up a fancy, folding, lean-back barber chair. Then Mr. Durant would come in through a side door for his morning shave. Cordially he would gesture and smile. While the barber was busy his secretary would come in several times with shorthand book and pencil, ask questions or take down orders, while we sat around as so many onlookers.

Often times it would be almost noon before it was our turn, in which case he might ask us to join him for lunch. This meant going down the elevator with him to the ground floor, hustling across the street to a lunch counter where we would have a sandwich and a glass of milk or coffee. Then, back to the office we would go, gambling as to whether we would get a check or a request to return the next day.

I will never forget Zeder's description of another such time when the very next day was our payday. Since we had only half enough funds to cover our payroll, it meant that we needed our check from Durant.

At Durant's office, before Fred had half a chance to say anything, Durant said, "Fred you are going out to the baseball game with us this afternoon. My wife is going to join us, and we all will drive out together."

Fred, in the rush of things, did not have time to phone us. As the game wore on, he kept trying to determine how to bring up the matter of the check gracefully, but never did succeed.

The next day we returned to Durant's office. Again there was the usual performance. Our office at Newark was on pins and needles waiting to hear the result, until suddenly Durant came across with a check, and we were able to meet our own payroll. Working with Durant did have its nervous moments.

Durant's future plans were indefinite, and contract funds were tapering off, so we decided to promote a car of our own. Since Dillon and Reed were the fast moving promotion brokers at the time, we made an appointment with young Dillon, and spent an afternoon in his office. He was interested and a good listener, but after a visit or two, decided it would not be timely to proceed on a new automobile venture.

Then our good friend and associate, Mr. Rayal Hodgkins, now retired as sales manager of Studebaker and living in the east, suggested that we go to Cleveland. (Ed. note: At this time, Hodgkins was manager of the Cleveland Tractor Co., per Automotive Industries.)

Cleveland was a logical place for another auto plant. The White Steamer had been successful there. Also, Jim Hesslet, whom we knew, had left the Studebaker plant and joined Rollin White who headed the Cleveland Tractor Company in making what became popularized as "Cle-Trac." Also, Rollin White had decided that there was a field for a smaller car, and was experimenting with what was to become known as the Rollin car. We did some consulting work on that car, but it never went into production.

Hodgkins and Fred spent a week in Cleveland making the rounds and promoting the idea of building a "Zeder" car in this Detroit's competitive city. For a time, definite plans were underway to build a new car called the Zeder, but nothing materialized of this venture. Fred and Hodge came back fully disappointed.

We Rejoin Chrysler Now at Maxwell

Meanwhile, our design was pretty well completed for a new car. An engine already was on the dynamometer and well along in the development stage. It was a smooth, quiet running engine with good performance. We decided to rekindle Mr. Chrysler's interest in the motor car business. Fred called Mr. Chrysler, and invited him over to see our new engine in operation. The date was set for a Saturday afternoon. All was arranged. Frank Schwartzenberger stood by to demonstrate the engine, Tobe Couture to build up the driving thrill.

Mrs. Breer and I had long planned our baby son's first trip to California for a proud visit with our friends and my relatives. It just so happened that we were scheduled to leave that very Saturday morning that Chrysler would visit, and our ticket reservations were all set. My first thought was to delay our trip for several days to join Skelt and Fred when they approached Chrysler. However, it was decided that was unnecessary.

Later they said that they had lunch with Chrysler at our favorite Newark eating place, the Robert Trent Hotel. Then, with Tobe as driver, they drove out to the laboratory at Elizabeth. Well, Schwartzenberger put on a swell show!

As I understand it, while Fred and Skelt reminisced about the past, about the electric sign of Chrysler on the Durant plant, they said, "Why not build a real car with our engine under the name of Chrysler?" Evidently they had built up his enthusiasm from the demonstrations they had given him to the extent that he put his arms around the both of them, and said enthusiastically, "We'll do it. Let's go." Those few words became the start of Chrysler Corporation.

Skelton said to Mr. Chrysler, "Naturally time is of the essence in our setup. When may we see you to discuss matters further?"

Mr. Chrysler replied, "Monday, but that is Labor Day."

"That makes no difference," was Skelt's answer.

Come Monday, Labor Day, Skelt phoned Mr. Chrysler reminding him of the Monday appointment. Skelt was taken aback at Mr. Chrysler's reaction to his call. Mr. Chrysler at first seemed to rebuff him, but soon settled down and said, "All right, you and Fred come over this afternoon."

They did. The discussion took place in Chrysler's family apartment on Madison Avenue, New York, and the groundwork was laid to go full speed ahead with the project. This was great news to me on the West Coast.

Back in Newark, things were quickly on the move. Designs were completed in detail, the engine thoroughly developed. Chassis and cars were built, and road testing was soon underway. Things on Mechanic Street were really humming!

Then came time to move to the Maxwell plant in Detroit and set up shop. Skelt and I were the first to go as advanced guard. We had our first meeting with Charlie Morgana at the Pontchartrain Hotel. Charlie was in charge of Maxwell production at the Highland Park Plant. We had visited the Maxwell plant and engineering group previously. Our view was that Maxwell engineering had lost control, or rather never had control, over the Maxwell product.

As was typical in most plants, production dictated the policy. For example, if John Doe on the test line felt that the valves or pistons were making a noise, he would contact Tom Brown on the valve guide or piston line and personally change the clearance limits. As a result you could hardly recognize the part from the detail drawing. This practice had to be stopped. Skelt and I met with Charles Morgana and his crew and stated emphatically that each part had to be made exactly as dimensioned and described in the drawings. After considerable argument, Morgana consented. Our instructions to him were that should irregularities or difficulties occur, or improvements suggested, a meeting would be held to decide whether to make the change. If everyone agreed, then the drawing would be changed. The date and details of the change would be recorded in the notice space in the right lower edge of the drawing and, in addition, change notices would be sent out. Only in this manner could engineering coordinate with production, and both have full control of the product. Mr. Morgana was hesitant at first, but later was enthusiastic to the point of calling this the greatest thing that ever happened to him and his men.

Our next move was to renew our acquaintance with the engineering organization at Maxwell. Skelt and I introduced ourselves to the desk man in the lobby of engineering. This man, Walter E. Mettler, was very courteous. He took us up to Mr. Freeman who then headed all Maxwell Engineering. There we met up with Howard Maynard and reviewed what engineering was doing. They were working on three different major car developments. One was headed by Claude Sauzedde, a French engineer, who was responsible for the complete design. Another was led by Emil Nelson, a former chief engineer of Hupmobile. A third was that of Freeman and Maynard who had their own ideas for new models. It was a three-corner competitive program, undoubtedly forced on central engineering by top management.

The Maxwell plant at the time was building a four cylinder car, as was the Chalmers plant. Both were on the borderline of horseless carriage era.

Our plan was to move our engineering group to Detroit, and set it up separately so as not to upset the existing engineering group at the Maxwell Plant in Highland Park, which was directing the engineering of both Chalmers and Maxwell plants. Chalmers was having difficulties. We were pretty certain that Chalmers' problems stemmed from their attempt to produce a larger, more expensive, overhead valve, six-cylinder automobile that was not thoroughly developed. As soon as a few had reached the field, they realized too late that they had a big problem on their hands. (This was not true of the four cylinder Maxwell which had a husky, two main bearing crankshaft, its two main bearings being of the large, preloaded ball bearing type. The four was rated quite high and was in demand in the used car market.)

As matters unfolded, the Zeder, Skelton, and Breer Engineering Company was not dissolved, but absorbed by the Maxwell Corporation. Our group would take full charge of all engineering. Maxwell production was to continue, and the Chalmers inventory was to taper off. We would be involved in designing and introducing a new Chrysler car. Engineering at Maxwell was not the least disturbed by our operation since we set up our headquarters at the Chalmers plant on Jefferson Avenue, some 11 miles away, on June 6, 1923.

An overview of the Maxwell plant in Highland Park in June 1923 where the Zeder-Skelton-Breer team first set up shop. In the background can be seen the Maxwell test track. (Breer Collection)

We Create Chrysler Corporation

One of the dynamometer test stands Breer found in the Maxwell plant. It was replaced within the year. (Breer Collection)

Our designers, equipment, and offices were settled on the upper floors. New dynamometer equipment and laboratories were installed on the ground floor. While our designers were refining details, releases, etc., the laboratory was checking and testing experimental engines and various assemblies. As soon as possible, sample cars were road tested experimentally while production was tooling up under the supervision of Billy Kilpatrick and his able assistants, Lou Beedy and Scotty Smith. Smith was a very fine engine man who cooperated with us 100 percent at Elizabeth. Vern Drum, a younger man, was Billy Kilpatrick's able helper along with the experienced background of the two able manufacturers, Frank Keller and Lou Beedy.

The Emergence of Four-wheel Hydraulic Brakes

Before leaving New York, Mr. Chrysler told us that the Detroit group had released some refinements for the current production Chalmers cars. Also, they had released for production a western development known as "Lockheed" hydraulic four-wheel brakes. They were to be standard on the Chalmers and would be in production about the time we arrived at the plant. He did not know much about the Lockheed brakes but said, "Better look into what it is all about."

There were two Lockheed brothers in southern California. One of them, Malcolm Lockheed, had a four-wheel hydraulic brake system for motor cars and contracted to adopt them to Chalmers cars on a royalty basis. (Malcolm's brother was the founder of the Lockheed Airplane Company.)

I found Malcolm Lockheed at the Chalmers plant following the details of the brake equipment on the assembly line. The brake system released consisted of four highly finished cylinders about 1-1/2" in inside diameter, open at each end, rigidly mounted on the axle parts, and located inside the overhanging wheel brake drums. Inside each cylinder were two opposing pistons with piston rods pushing outward. Face to face in front of the pistons were "rawhide" cups shaped like the kind used in bicycle pumps; that is, spring fingers pressed the rawhide cup lips against the cylinder surface to seal against hydraulic pressure leakage of liquid when applying the brakes. Held fast in the cylinders were two opposing coil springs. When the brake pedal was released, they would push the pistons inward toward each other forcing the brake fluid back from all four brakes through flexible hose connections and copper tubing to a single master cylinder which had a piston directly linked to the foot brake pedal. If a slight leakage occurred, the brake fluid could be replenished by supply tank through a valve mounted on the dash. Small bleed valves at each cylinder would drop out all air when the system was filled. (Air in the system would act as a cushion, and spoil the feel of solid brake action.) Flexible rubber bellows snapped into place over each cylinder end to keep the road dirt away from the exposed cylinder surfaces. Finally, the brake shoes were rigid, ribbed castings with brake linings riveted to their surfaces. The two shoes were hinged on a common pin at the bottom, and spring-pulled toward each other to free the shoes from the brake drum.

The fluid used for production and service was to be known as Lockheed brake fluid. It consisted of glycerine and alcohol, the object being a fluid that would seal, lubricate, and yet not freeze in the coldest weather.

In my discussion with Malcolm Lockheed following his detailed description of his system, I asked many questions. Being a Californian, I knew about the long grades and hot temperatures found in the desert. I asked Malcolm how the alcohol, which boils at fairly low temperatures, would behave on the desert. On a long grade I surmised that the combination of the high atmospheric temperature boosted with the heat of brake friction would cause the alcohol to build up sufficient vapor pressure to keep the brake applied even after the driver took his foot off the brake pedal. Malcolm replied enthusiastically that he had been all over the desert, thoroughly tested the brakes under all conditions, and had run into no trouble whatsoever. He seemed so overwhelmingly positive that I became suspicious of his statements, and decided we had better find out for ourselves. I took our boys in tow and said, "We have no time to go to the desert and we have no hills in this vicinity. Let's take a car out this afternoon while the day is still warm and drive it as long as we can with the engine pulling wide open in gear and the brakes on to simulate going down a long desert grade to see what happens." We drove along Windmill Pointe Drive, a two-way drive with a wide parkway between allowing ample room to turn back without stopping. Before long, everything let loose. The brakes let go all of a sudden without warning. Brakes and alcohol smelled in every direction. Things were steaming hot.

We went back to the plant, only a few blocks away. When we took the brakes apart, we learned a lot. The vapor pressure question was secondary to what we found. The rawhide cups had all shriveled up hard and dried like bacon fried in a pan. Before curing, rawhide, by nature, grows at fairly constant animal temperature, therefore never has to contend with temperatures above that of the atmosphere or never to freezing temperatures. Fortunately, we discovered this problem before too many cars had come off the assembly line. It called for quick action! We liked the basic idea of hydraulic application because it meant equal brake pressure at all four wheels regardless of the wheel jumping up and down over road inequalities.

This was just the opposite of mechanical four-wheel brakes that had become popular at that time and were well advertised. Mechanical brakes meant a system of rods, equalizers, cross-shafts in bearings, all connected to the foot brake pedal which was attempting to distribute its forces uniformly to all four wheels while bouncing up and down. Aside from the many exposed parts, the thing that we did not like about them was that friction due to vehicle motion, especially under the high pressure of brake application, would interfere with obtaining equal brake pressure at all four wheels. So we were all for the hydraulics and their opportunity to make good.

I immediately got Mr. Bockins and Johnnie Merrell of the Manhattan Rubber Company on the phone and asked them to come out and discuss our problem. Manhattan Rubber had built up their business in the high temperature, high pressure steam sheet packing field. They had a background in various types of rubber compounds, both with and without fabric reinforcement. Also, they were in the field of making special rubber molded shapes. To our thinking, they were the best source to develop dependable rubber cups to replace the rawhide quickly with our guidance.

It was a question of day and night action. We knew glycerine was not a preservative for rubber. After discussing many compounds, we found that castor oil was a base since it was a preservative of rubber, but as to alcohol as a thinner, which we thought best, we had no information. There was no time to lose on long-time experiments. It just so happened that I knew the top chemist at Parke-Davis, Mr. Briggs, who did a library search and located a small paragraph in one of their basic books that stated that alcohol was classified as rubber preservative.

So we were on our way. Overnight we designed and developed various rubber cups, with and without fabric reinforcing, and various lip edges to best seal them from loss of liquid. Our developments started with bronze-stamped, cup-shaped fingers to insure lip contact against the cylinder at all times — winter or summer.

The importance of perfect lip edge functioning is that every time the brakes are applied, the combined length of all the cup lip edges moving back and forth added up to around 40 inches, like moving a squeegee with a lip edge 40 inches long under all types of pressure and temperature without losing hydraulic liquid from leakage or seepage. The bronze lip pressure fingers were soon replaced by a simpler, more effective, and dependable coil spring that pushed the lip edge outward against the cylinder wall.

Through the cooperation of our new production organization and the Manhattan Rubber Company, we soon had the correct material coming in and being installed in cars on the production line.

Howard Maynard had one more concern that was bothering us. Over a period of weeks, various experimental cars equipped with the Lockheed equipment would lose their brakes without warning. It was due to air getting into the system, air being compressible whereas brake fluid was not. There was no prediction for when this would happen. Sometimes it would not take place for weeks, then it would occur on the first day of operation.

I well remember the Saturday that Howard and I spent the day checking cars just off the line which were to be shipped the following Monday. We had a late lunch at the Chalmers plant, and while eating and discussing our problems, it suddenly came to me what was causing these troublesome failures. In fact, I jokingly said to Howard, "I'll bet you a dollar that I can go out and in a few moments put every one of those hydraulic brake jobs out of commission. Let's go out, and I'll show you." As I expected, all I had to do was to push and pull the brake pedals back several times and I had air in the system. What happened was that when I applied the brakes, the push on the brake pedal forced the fluid to all brake cylinders, pushing out the pistons, compressing the opposing springs and applying pressure to the brake shoe. On my release of the brake pedal, these compressed springs would reverse the fluid flow to the master cylinder, forcing the brake pedal back to starting position. The time it took to move the fluid and pedal back gave the feeling of sluggish or slow brake release. To overcome this feeling, the design division had added a spring to help the brake pedal come back faster. At some critical times, the pull back spring and foot action would be enough to pull air in past the master cylinder cup similar in action to the familiar bicycle pump.

That very afternoon the two of us drove over to the experimental shop at Maxwell and made a one-way slip connecting link to replace the solid link between the brake pedal and the master piston. On Monday, all cars were shipped with the new slip links. It was the answer. The pedal would come back independently by spring action quicker than one could move one's foot. The sluggish feeling completely disappeared. We relaxed, knowing that air could never get into the fluid side of the system as long as the fluid pressure always remained above that of the atmosphere.

Selling the public on four-wheel hydraulic brakes, however, was a rough go. Competitive advertising as well as insinuating verbal salesmen's expressions were vicious at times. A typical example was that of competition comparing our four-wheel hydraulic brakes to the mechanical. The only hydraulic brakes used on automobiles before were on Duesenberg which was five times higher priced than Chrysler. They were of a different design, and only a relatively few cars were built. Four-wheel brakes being then built in quantity were all mechanical. Incidentally, at this same time, Mr. Erskine of Studebaker tried to get a group of manufacturers to stay with two-wheel brakes, and advertise opposition to four-wheel brakes, claiming that two-wheel brakes were a lot safer.

One case in point involved one of our largest competitors, with whom we competed in the lower price bracket with our four-cylinder cars. They had a display spread hanging on the wall of a large salesroom. All the parts that constituted their mechanical brakes — rods, equalizers, cross-shafts, pins, and even the cotter pins were in one group. Next to this were all the parts of our hydraulic system — cylinders, tubing, rubber hose, pistons, rubber cups, etc. After explaining all the advantages of his mechanical brakes as he saw them, the salesman then would insinuate that he didn't think that anyone would like to use syringes for a braking system.

By the time we developed what we considered to be good commercial hydraulic brakes, our patent coverage was such that we could have featured our developments as Chrysler hydraulic brakes. There was practically nothing left of the Lockheed system beyond the idea of using hydraulic fluid in place of metal rods, etc. As a result, we agreed with Lockheed through its representative, a Mr. Scott, to give them all our patents in return for a license free from all royalties. This pleased Messrs. Scott and Lockheed, and gave them a chance to break down opposing barriers by selling hydraulic brakes to the motor car industry one by one, thereby converting them to hydraulic boosters. It also allowed them to put their name on hydraulic fluid on all garage shelves and oil stations, and thereby stop uninformed service mechanics from substituting detrimental lubricating oil. So the world now drives in safety on very dependable hydraulic brakes.

The New Chrysler Six Emerges

Even though we were working on the Chalmers production cars, there was no interruption in our development of the coming new Chrysler. Frank Schwartzenberger was operating our dynamometer day and night on testing. Skelton was busy completing all design details, releases, etc. There were last minute decisions and releases to be made, while important major assemblies had to be tested both in the laboratory as well as on the road.

Also there were a number of new innovations or features added for the first time with the introduction of the Chrysler car. For example, one day, while we were still in the east, Mr. Chrysler said to the three of us while we were in his office, "There was a fellow in here yesterday by the name of Mr. Abeles who tried to interest me in an oil filter developed by a western inventor, a Mr. Sweetland. While you are here and have spare time, I wish you would take a look at it. Guy Nonemaker will call them and tell them you are coming over."

We found Mr. Sweetland to be an experienced filter man who was very successful in the Mother Lode Mining district of California building and installing large filter systems used in gold refining. The small filter he showed us demonstrated very well. Dirty black crankcase oil that passed through the filtering plates was seen coming out of a nozzle inside a glass tube crystal clear, the oil still in its natural, clear, amber color.

This was something new for the motor car, and was a worthwhile feature for two counts: It would definitely remove the accumulation of wear particles in the crankcase, and with a sight feed on the dash, the car owner could see that it was doing its job. This became another first for our new car.

Skelt and our design division took on the responsibility of making the filter into a commercial unit with replaceable filter elements and providing the oil sight feed. The system became known as the "oil by-pass filter system" wherein only the excess by-pass oil went to the filter. This has been standard for many years and has gradually evolved into the full flow filters now used. "Full flow" means that all the oil that goes to the engine bearings must first pass through the filter.

Road dust was something we continually worried about. Having decided to filter the oil, we felt it was just as important to remove all dirt possible from the air entering the carburetor. At the time there was a rather ingenious all-mechanical, rotating, self-cleaning, and non-clogging centrifugal dry type air cleaner made in Chicago by the United Air Cleaner Company. As the air came down over a vaned, self-propelled, free running cone-shaped stamping, it forced the heavier dust particles out away from the air stream as the air quickly turned under and inward to enter the carburetor.

It was quite effective at the time, and has since evolved first into the combination intake muffler and oil type filter, then later to the present, impregnated, replaceable porous paper filter that is standard equipment on all automobiles.

Next we started the vogue of using a symbolic ornamental radiator cap as standard equipment in place of the traditional plain caps. We used the deity wings of Mercury in our design to designate speed. Moto-Meter, a popular, ornamental radiator cap with a thermometer column which could be observed by the driver, was widely used by everyone at the time. To avoid disfiguring our Chrysler Mercury wings, we placed the temperature gage on our instrument panel where it could easily be seen. Others began to follow, making the Moto-Meter obsolete. George Townsend, president and owner of the Moto-Meter Company, who was a close friend of Walter Chrysler, realized that it would put him out of business, yet he was a very good sport about the change.

While at Elizabeth, New Jersey, we did a lot of development work on cylinder head design. In order to get breathing capacity for higher engine output, we wanted as much room as possible over the "L" head valves. At the same time we wanted a high engine compression without so-called spark knock, basically known as detonation. We worked on the theory that when the charge was fired, the rapid progression of the flame over the piston area of the combustion chamber could be slowed by bringing as much area of the cylinder head down close as possible to the piston at the top of its stroke. Then we would rib the adjacent head surface to offer more cooling area, thus slowing the rapid flame travel, and thereby reducing detonation.

We made several heads, some with finger grooves and one with no grooves at all. The flat head gave us the best results. This cylinder head in combination with a seven-bearing crankshaft fulfilled all our test requirements for an engine that would stand 3,000 rpm at wide open throttle for 50 hours. We were proud of our engine. It delivered an authentic 68 hp with the 50-55 octane fuel then on the market, and with a 4.7:1 compression ratio. (Ethyl gas came on the market several years later.) We thus had a fairly light car, around 2,700 lb, with standard 56-in tread that would do close to 75 mph.

Finally it came time for Chrysler to show our first model cars in January, 1924, at the time of the opening of the New York automobile show. It happened at the last minute. The auto show people according to their ruling would not allow a Chrysler to appear on the floor because the car had not been in production for a given period of time.

This rare photo, dated February 3, 1925, shows Maxwell and Chrysler still as separate entities. The Chrysler car became so successful that Maxwell was absorbed into the Chrysler camp by reorganization in June 1925. Driver of the car is Ralph DePalma, popular and successful race car driver of that day who campaigned the first Chryslers around the nation in road races and other events. (Breer Collection)

Skelton, Breer, and Zeder pose with the original 1924 Chrysler Phaeton in its restored shape. (Breer Collection)

Chrysler requested Joe Fields, then sales manager, to engage the lobby of the Commodore Hotel for a private show. It was a great success, more public than private. The car was featured as a medium-priced automobile with a high power, high speed, high compression engine, four-wheel hydraulic brakes, air filter, air cleaner, etc., a lot more features than were in high priced cars or to be had in others at a medium price. (Ed. note: Breer is in error. Chrysler had several cars in the show in the Chalmers exhibit area, and did display a roadster in the Commodore lobby with several other exhibitors.)

Some 32,000 Chrysler cars were built the first year, 1924. People liked them because they could drive as fast as they dared, yet the car would stay right with them, and seemed to ask for more. This was something most cars of that time would not do. The buying public put their stamp of approval on our product in spite of the fact that competitive salesmen were vociferous in claiming that people would not want to have one of those hybrid Indianapolis racing engines in their cars. It took a year, through advertising and publicity, to educate the public into understanding that our engines were not turning over any more revolutions per mile of car travel than the average car on the market. In fact, we pointed out that the engines in some competitive cars were being asked to turn over many more revolutions each mile than ours. This stopped it.

Shown here are a pair of 1924 Maxwells, the one above pictured at some undetermined curbside in Detroit, the one below outside of the Statler Hotel in downtown Detroit. The success of the first Chrysler was so phenomenal that by mid-1925 the Maxwell-Chalmers organization had been reconstituted as the Chrysler Corporation and the Maxwell car revamped as a low line Chrysler model for 1926. (Breer Collection)

Ken Lee, head of Research under Breer, poses beside an experimental small size car under test on Michigan roads. It had a unique radial engine and front wheel drive, both quite radical for that time. The project was dropped because it was not considered commercial enough . . . except for the engine's crankshaft which was picked up by Air Temp and used in its air conditioning units. The radial engine is shown below.
(Breer Collection)

Thanks to Mr. Chrysler's understanding and cooperation, Chrysler engineering also continued to grow. He never questioned our judgment in spending money for added dynamometers, laboratory or road testing equipment, or added design facilities. As a matter of fact, we were so preoccupied with our engineering that we neglected our personal well-being like many others, and when the market crashed in 1929, we came into a financial jam. Without hesitation, Mr. Chrysler came to our rescue.

Shortly after the market broke in 1929 he called a meeting and issued an ultimatum that the head of every division had to make an overall cut of 20 percent in expenses, wages, employment, etc. After the meeting was over, and we were walking with him to our offices, I made the passing remark, "Mr. Chrysler, does this mean we have to cut out some of our engineering research or laboratory projects?" He put his arm around me and said, "Hell no, go right ahead. Just don't say anything about it."

Chrysler never did fail to support the engineering end of our business, and demonstrated this on numerous occasions, one being when we were exploring a small, lightweight car of a radical nature — front-wheel drive with a five-cylinder, air cooled radial engine facing forward. It was one of the projects underway in our research division headed by Ken Lee. On one of Chrysler's visits to Research, I mentioned that there were times when we got a bit off the beaten path, and I hoped that when we showed him this specific project he would not think it too radical. Even before he saw the car he put his arms around my shoulders and said, "Don't you ever think that. I love it!" He later expressed much interest in the project and always passed along encouraging remarks. Although the automotive application for this design didn't prove practical, we eventually were able to use the unique crankshaft design for air conditioning in our Air Temperature Division.

Part IV: Reminiscences of Early Product Developments at Chrysler Corporation

The Chilled Cast Iron Valve Tappet

We never relaxed in exploring engine improvements or adding overall refinements and endurance life. Our ambition always had been to have a perfect motor car on the market, a car that would run several hundred thousand miles with no more than ordinary attention. What we had in mind was to continuously obsolete our past models by introducing innovations with more attractive and desirable features. This is really what the automobile market came to be many years later.

That our engines were able to run at full power on the test stand for 50 hours at 3000 rpm with the throttle wide open was only the foundation on which we continued to build. We were always alert to find new ways to increase the material life of our flat-face tappets and the cam surfaces that operate the opening and closing of the poppet valves, recognizing higher operating pressure as engine speed went upward. In spite of using the finest of steels and the finest heat treatment for maximum hardness of surfaces, scuffing or breaking down of the surfaces would occur, leading to noise and failure. One day a Mr. Wilcox of the Wilcox-Rich Company at Saginaw, Michigan, came in with an experimental tappet design that looked very attractive. His tappets were a two-piece design: a steel upper sleeve with threads at the top for the valve clearance adjustment screw, and a chilled, cast iron, flat-faced mushroomed platform that rode the actuating cam for the lower end. The steel sleeve was pressed onto the mushroom stem tight against a shoulder. The sleeve was then electrically spot welded at five or six places around the side, making it a unit structure. The unique part of the chilled cast iron face was that its face surface was almost diamond hard. The sudden chilling of the molten cast iron caused vertical crystals of some length to be formed backing up the surface face structure against the high impact forces of the operating cams.

The technique of using two entirely different hardened metals working together was entirely new and proved to be a godsend to automobile engines. To this day no material substitutes have been developed or found. Now camshafts are made of cast iron with chilled face cams, and the tappets have hardened steel faces. Thus roller tappets were forced into the background, never to appear again on the production line.

Pioneering a new tappet design was fine, but we soon ran into trouble between our shops and Mr. Wilcox, our supplier. Mr. Wilcox was set in his ways and determined not to allow anyone but his pledged workmen to enter his plant so that he could keep his manufacturing process a secret. Unfortunately, the cast iron head of the two-piece tappet sometimes would come loose during a production engine test. With no apparent manufacturing change, more and more were coming loose. It was not long before our production division was up in arms and reported to Fred Zeder that they were going back to one-piece steel tappets because the Wilcox tappet was too much trouble. They had come to the conclusion that the tappet was not commercial. This was on a Friday when Fred came back to Engineering and told us about the problem. Production was to hold a meeting the following day, Saturday, the outcome of which everyone knew would be to tell Mr. Wilcox that we were through with his product. I said to Fred, "They are all wet. If we allow production to go back to steel tappets we will have a hell of a time to get them back to the one thing we cannot do without — chilled cast iron tappet faces. We better go over, sit in the meeting, and see what can be done."

That Saturday, Fred, Skelt, and I went over. There were about 40 people sitting in on the session. Mr. Wilcox and his son Merrill were present as well as all of our production men. There was not a smile in the crowd.

Samples were shown; strength tests were made; chilled faces were broken — all to show the inconsistency of the product. The battle was on. Production wanted to return to all steel. Naturally we as engineers pleaded tolerance. I said to Fred, "I think there is a way out, at least for another try." I asked Mr. Wilcox to allow our Theodore Koerner and an assistant or two to go to his Saginaw plant and work with his production men to lick the troubles. I said that we want this tappet, if it were possible through Mr. Koerner's cooperation to produce a good product. Mr. Wilcox realized what he was up against and at last decided to open his plant to our men. We did resolve the problem, and after that, thanks to Mr. Koerner's fine accomplishment, our production line was never again shut down for tappet difficulties.

Mr. Wilcox was very appreciative of what we had accomplished for him and we never lost his friendly association with us. He was an elderly man, and when he lost his son Merrill, who also was his co-partner, through a bad accident, he decided to get out of the business. The brokerage firm headed by Carl Higbie had an eye on the plant, and worked up a deal for taking it over through his contact man, Mr. Carl Flinterman. Mr. Wilcox would not close the deal without the three of us placing our stamp of approval on the transaction. Finally, I drove to Flint with Carl Flinterman to give our approval, and the deal was sealed with a $25,000 check Flinterman handed to Mr. Wilcox. Let me add that none of the three of us had any financial doings in this deal. In fact, it was our practice never to invest in any of the supplier companies with whom we were working.

Revamping the Maxwell Four for Power and Endurance

At the first opportunity we had when the dynamometer was clear after the Chrysler Six went in production, I suggested to our dynamometer crew that we run off some power curves to find out what the Maxwell Four, then in production, could really do. It took us two weeks to make a characteristic power curve that should not have taken over a day. What happened was at wide open power pull at a given speed, the scale beam would raise up to a maximum, then fade away. The exhaust valves would get so hot after a few runs that they had to be ground (reseated). This took time.

The engine had a separate intake valve port for each cylinder (we never could understand why). The two center cylinder exhaust ports were siamesed; that is, two adjacent cylinder valves exhausted through a common port connected to the exhaust manifold. What we found was that although the intake valves had water cooling completely around the valve seats, the siamese twin exhaust valves in the common port had water cooling only part way. When the engine was pushed too hard, the valves would become red hot, valves and seats would warp, and the engine would start to misfire. When this happened, the driver naturally would slow his pace. Then the valves would cool, and the engine would settle down again to cure the problem. All we had to do was to make new exhaust and intake manifolds, and change the camshaft so that the intake and exhaust valves were switched. The intake valves now became the exhaust valves, properly cooled, and the cool running twin exhaust valves and port became the intake valves. In addition, we made one other important major change, to convert it to a three main bearing engine instead of a two bearing type. (Ed. note: This would have been the Four used in the 1927 Series 50.)

With these changes and other minor refinements, it was not long before we released "The Good Maxwell" four-cylinder engine.

Later we designed an entirely new four of a more modern design. It had split half main bearings with a crankshaft flange to hold the flywheel, and a pressed steel under pan for the oil and complete enclosure. As an "L" head design, it was more sturdy than the "Good Maxwell" four, and had a full pressure oil system and greatly added range of power and endurance. We introduced it as the Chrysler Four. When we added rubber engine mountings to it, the Chrysler Four put the competition to shame with its smoothness. In fact, one could hardly tell the difference in normal driving between it and the Six.

The Development of Aluminum Pistons

From the very beginning we had opted in favor of die cast aluminum, split skirt pistons developed in cooperation with Mr. Frank Jardine who was with the Aluminum Company of America. The Jardine piston had a split skirt, and was cut and shaped with varied wall thickness to balance between a minimum wall clearance and not too much wall pressure, thereby avoiding piston scuffing by rubbing friction.

There were two points in favor of and one against the use of aluminum in place of cast iron. Aluminum was much lighter than iron, and cut down the tremendous forces required to reciprocate the pistons up and down. It also was a much better conductor of heat to the cylinder walls. The one factor against aluminum was its excessive temperature expansion which required greater clearance between piston and cylinder wall. It was in attempting to cure the latter that we adopted the invar strut, invar having zero expansion.

The Aluminum Company of America, developer of the aluminum piston, was the only source we had with a background in die casting and heat treating dependable aluminum pistons. They had surrounded their work with patents, and their price per piston carried an ample protection profit so that they could battle the elements through the courts on patent procedure.

Also at this time, Charles Bohn of Bohn Aluminum was trying to break into the aluminum piston casting business. Naturally the high price per piece made it very attractive. Bohn Aluminum, as I remember at the time, had made dies to cast some pistons for the replacement service trade.

We desired our pistons to utilize the experimental Invarstrut concept but the Aluminum Company of America showed no interest. Bohn Aluminum was close by and anxious to work with us, so we began a relationship with them through which we experimented with this new technology. It was a choice made with some apprehension.

Aluminum was a mystery metal, and the Aluminum Company of America had much background from years of pioneering. This would make them a more dependable source for production than Bohn. There was a sense of uncertainty in those days about heat treating aluminum die castings properly so that the metal would not grow or warp with age. But at the same time we liked the way Bohn would cooperate even though we had doubts about their experience and ability as a production source.

I went all out to get the two companies to work together. With all arguments spread out before them indicating how favorable it would be to cross license with respect to piston work, we simply could not induce them to amalgamate.

Finally, we gave the business to Bohn, and asked our metallurgist to work closely with them. We began by installing Invar-type pistons in one of our top engines, but it was not too long before it developed some loose struts. (Ed. note: This probably was the 1928 Chrysler Series 80.) The engineer on the Bohn job was a Mr. Nelson, who of his own accord had been working on a steel strut in place of invar for partial expansion control which by its nature worked in opposition to the aluminum expansion, retarding it, so to speak. With this he provided a continuous aluminum one-piece structure — aluminum for strength and steel added for expansion control only.

Now we found ourselves with two aluminum piston design ideas. In one sense we favored our open-type structure Invarstrut piston design having perfect expansion control, while Nelson leaned toward his steel insert design which still required a partial split skirt avoiding the weaker split skirt in Jardine's Aluminum Company of America piston.

We liked the Invar pistons because when the struts were held physically tight, they showed perfect expansion, allowing us to operate with less clearance than with a cast iron piston. Endurance tests showed very uniform bearing contact between the cross head type of piston skirts. The piston head carrying the rings was free to expand uniformly of its own accord. At the time we had to depend entirely on the contraction of the aluminum during casting which resulted in a tight grip around the key-shaped edges formed in strut stampings. Today it would be possible and practical to cast with a strong and solid metal-to-metal bond between Invar and aluminum. Had we not been so busy otherwise at that time, we probably would have followed through on this development with complete success. But we did not, and Nelson's compromise pistons won out. In turn Aluminum Company of America lost out as a source. Sometime later other aluminum pistons were developed satisfying our requirements, and our production division set up a casting department to make our own pistons, saving us several dollars per engine built.

Pioneering the Downdraft Carburetor

One day in 1928 Everett Sheppard, an early development engineer with Holley Carburetor Company, came in and said, "I would like to discuss some downdraft carburetor ideas with you." All carburetors then were updraft except for a few horizontal or side draft types.

A downdraft carburetor concept would be ideal because gravity forces would accelerate rather than oppose the flow of liquid fuel particles, and they would be distributed better with less air velocity, producing more power. There would be little or no pockets for fuel to accumulate. My only question concerned cold engine starting. What would happen if an engine over-choked? We certainly would have difficulty starting it, once flooded, especially if it were an "L" head engine which had a reservoir capacity that conceivably could half fill the manifold with liquid fuel. (This we later resolved by placing an automatic drain check valve at the lowest spot in our manifold.)

Sheppard had a prototype that he had been developing on a laboratory flow stand. He was attracted to the idea of using a flexible diaphragm for the gasoline feed in place of the gasoline reservoir float chamber still universally used today. The diaphragm idea looked attractive to us because of its compactness and casting simplicity.

We had hoped to have the downdraft carburetor perfected for the 1929 models. Our first spot checks looked very encouraging, but Shep's diaphragm idea complicated carburetion clean up. It also brought in variables and unknowns. To be certain we would not be caught the last minute, I said to Shep, "Would it be agreeable if we brought Stromberg in to develop a downdraft carburetor using a float chamber as insurance in case the diaphragm idea did not work out?" He agreed, which was a good thing because Shep never did resolve the diaphragm riddles, and we finally had to release the Stromberg. It was only a matter of time before all other car makers turned to downdraft carburetors, and I doubt if you can find any quantity car producer today that is not producing vehicles with downdraft carburetors.

Reminiscences of Early Product Development at Chrysler Corporation   97

Battle of the Vacuum Tanks

Most people have forgotten that the engine carburetor once was supplied fuel by gravity from a quart-size tank located on the dash under the hood. At that time updraft carburetors were standard equipment. A trip trigger mechanism operated by a float pulled the gasoline from the rear tank by means of the engine vacuum, alternately filling the spill chamber portion of the tank. This system was popular on all cars with the exception of Ford. On the Model T the gasoline tank was mounted on the top of the dash in the engine compartment. This way Henry Ford saved the cost of a vacuum tank. But in the long run, he had to give it up because a head-on collision might break the gravity line to the carburetor with undesirable consequences.

Stewart Warner was the primary source for the vacuum tank patented by Webb Jay. His high profit monopoly price induced others to attempt to break in. Bill Sparks of the Spark Whitigan Company of Jackson approached us with a vacuum tank much cheaper and better and, he claimed, free of patents. However, Stewart Warner stopped him by a patent law suit. Others had tried too and were stopped. The high profit margin induced the Albert Champion Spark Plug Company at Flint, Michigan to develop a camshaft-operated gasoline diaphragm pump. However, it would require changes to the engine block and the addition of an eccentric cam on the camshaft. In its favor it was low cost, and provided constant gasoline flow.

While Champion's development was in process, in came a Mr. Naylor who had the backing of the American Can Company. Naylor had developed a vacuum tank that he could sell to us at half the unit price of Stewart Warner's tank. I believe the Naylor tank was a European development. The tank passed all of our tests and we placed an order with Naylor contingent upon his tank being free from all patent litigation.

After we released it for production, in came Webb Jay one morning from Stewart Warner to warn us that his company would issue an injunction against Naylor for making the tank which would be embarrassing to us as well as tie up our production.

"Webb," I said, "Naylor is backed by the American Can Company. He assured us by written guarantee that our company would be entirely free of any patent law suits. You know, Webb," I continued, "Stewart Warner has insisted on a high price with ample cushion to cover patent lawsuits. As a result, they have virtually invited everyone to compete. Webb, you know that this tank has been built for years in an expensive manner so that it can be disassembled for service purposes, but the service requirements are nil or negligible. Why don't you build a sealed tank on the American Company's automatic can machine, install your mechanism in a simple manner, and get the price down to one dollar. This way all you have to do when a customer comes in for service is rip the tank open with a can-opener. If its defective, give him a new one, no charge! If its okay, show him why, and charge for the new one. Gets the price down where it belongs."

Webb thought for a minute and replied, "Let me go back to Chicago tonight and discuss it with the higher ups. I will report to you next week." But Stewart Warner could not be budged. As a result, we went with the Naylor tank which went into production without any interference. (Ed. note: It was introduced on the 1930 Chrysler CJ.) Later the gasoline pump came into popularity, and today few people remember or have knowledge of a vacuum tank system.

Breaking the Radiator Monopoly

Radiator manufacturers usually asked us to give them the size and details of our engine. Then they would design a radiator of the proper cooling capacity.

When we moved into Detroit, we discovered that the Maxwell Division had established its own fin and tube type radiator department which produced radiators for Maxwell. We knew from our past working with Studebaker that the new technique of cellular radiation was more economical than the fin and tube type. If we wanted to continue making radiators, we decided that we might as well set up a laboratory for radiator development since, to us engineers, there was a bit of mystery associated with the radiator business. In reality, it was something of a hangover from the earlier days of doing business by suppliers.

We brought in a brilliant young engineer, Nick Diamant, and set him up with a couple of tool makers and punch presses. Nick built himself a miniature wind tunnel by which he could ingeniously explore heat transfer through air flow within a single cell of a multi-cell type radiator core. After much experimentation, Diamant came up with a cellular type design that had a 30-40% increase in efficiency. We were proud of our accomplishment, and asked Fedders if they would make a full-size core from our design to cool our low priced car. After much delay, with hardly time enough to check it out for production, in came the radiator except with a much larger cell size than we had specified, the result being that the sample had insufficient radiation capacity. Naturally we had to release the previous production cellular radiator to meet our car schedule. To us, this was unfair play, so the following year we brought in an outside radiator manufacturer

(Jamestown Metal Equipment Company) who was not a part of the radiator guild. They cooperated one hundred percent and, with our design fully tested, took a large share of business away from Fedders who, in turn, loudly proclaimed that Jamestown would go broke because it could not make radiators at such a low price. They even took Mr. Diamant to court, accusing him of copying or stealing Fedders' design. The case was finally dismissed on its merit.

One day Lou Fedders and Clarence Batchelor, his Detroit representative, came in to see us to find out why we had taken our radiator business away from them, especially since they realized that Jamestown not only was not going out of business, but was expanding. We told them that their production head was too dominant, would not listen to his engineers, and they would be better off without him. Fedders went back to Buffalo and cleaned house. Later, he personally thanked us for enlightening him as to what was going on!

Our radiator laboratory in fact broke the so-called radiator price guild as far as we were concerned and opened up competition. Before, the various radiator manufacturers would respect each other's bids at various auto plants, eliminating any bargaining. This one move saved us between two or three dollars a car in cost—quite an item with our large production.

Our successful radiator efforts made obsolete the fin and tube manufacturing equipment used for building radiators for Maxwell cars. The surplus equipment was sold to Fred M. Young who used them to found the Young Radiator Company, which later became an important equipment source in the heat exchanger field.

The Engine Bearing Battle

One day Carl Johnson, founder of the Cleveland Graphite Bronze Company, came in and said, "I think I can save you some money!" This was a welcome comment, especially coming from a vendor. At that time all engine crankshaft and camshaft bearings were made to size of a heavy bronze shell lined with babbitt. The main bearings were split in halves, but the camshaft bearings were of thinner, non-split, cylindrical bushings. To slide the camshaft into the engine, the bearing support holes had to be large enough to allow the shaft with projecting cams to be inserted through the several crankcase bearing holes to get it into place. The inside bearing sleeves, already pressed into place, naturally were of a large diameter. This large outside diameter bronze shell with the small babbitt-lined hole made the bearing heavy and costly.

Carl Johnson's samples were of cast iron with a thin bronze liner pressed in for a bearing. The thin bronze bushing, Carl's original invention, was made of sheet bronze with die-pressed pits in the surface which were filled with a hardened graphite paste. We ran many tests on the bearing and found it very satisfactory, so we released it for production to replace the more costly, all bronze, babbitt-lined unit. We found out later it never made it into production because the regular supplier, Johnson's competition, who furnished the complete set of engine bearings, would underprice this one bearing below that of the Johnson unit. It was two years before we could break this practice which also resulted in one of the corporation purchasing men losing his job. The Cleveland Company now is one of the big bearing suppliers.

Introduction of Floating Power

Immediately after the introduction of the first Chrysler car, we concentrated on further refining what was already a good, acceptable automobile. The original engine in the car had four-point suspension; that is, the engine was bolted to the steel frame at its four corners, thus adding torsional stiffness to the frame. Chassis frames were quite simple at the start, but gradually evolved into box section side and cross members designed to make the frame torsionally rigid. By bolting the engine to the frame, engine noises were transmitted into the car body.

One particular noise that bothered us was the torsional vibration of the crankshaft at certain critical car speeds. We took care of this by adding what we called a "crankshaft impulse neutralizer" or what we described previously as a crankshaft torsional dampener.

Next we bolted the engines to the frame using rubber as an insulator. The engine was now held fast by sandwiching the supporting frame between two molded rubber cushion pads that were held in compression between an outer plate and the engine by means of through bolts. This design was quite effective in damping the engine noise. It also absorbed or deadened some of the higher frequency engine vibrations.

The combination of crankshaft impulse neutralizer and enclosed type rubber engine mountings made a marked improvement in our 1925 cars, the second year we were in production. By continued experimentation toward building smoother and quieter motor cars, we then discovered that putting rubber in compression was an improvement but in the wrong direction. Theoretically the enclosed rubber design could only be effective in cushioning vibration forces if the pressure on the rubber exceeded that developed by the fastening bolts. It's like a person leaning against a closed door when a person on the other side tries to open the door. He must overcome the force already against the door before it will move or open. We next eliminated the clamping pressure to allow the engine to be suspended as freely as possible. We felt we could do this by bonding the rubber to steel plates that would be fastened to the engine and the car frame respectively. Rubber tire manufacturers seemed reluctant to give this approach a try so there was an unnecessary delay in getting samples. Jim Zeder, in charge of laboratory work at the time, suggested that we form a rubber laboratory of our own. We found the man we wanted in a Mr. W. J. McCortney, who set up a complete rubber mill with molding equipment, and we soon were on our way molding experimental parts and making rapid progress in bonding rubber to metal, reaching a point where the bond had greater strength than the rubber body to which it was attached.

It was natural that tire manufacturers would have little faith or interest in bonding rubber to steel because of their earlier bad experience in bonding solid rubber truck tires onto steel replaceable rims. In fact, when I was superintendent of the Moreland Distillate Truck Company in Los Angeles, with an express service of some 60 miles between L.A. and San Bernardino, it was not uncommon for one round trip to call for a new set of tires because of the rubber peeling loose from the steel.

However, our research and development work in our rubber laboratory proved that a bond could be produced consistently that was stronger than the rubber body adjacent to it. In other words, the bond was so good that the rubber would tear rather than pull loose. To my knowledge I have never experienced any complaint due to a failure of the rubber mounting. In fact, in the early days, we often would demonstrate the strength of the bond by suspending a complete car via the rubber mountings. We also discovered that rubber life was improved by this activity.

While this work was in process, an inventor from the midwest came in with a car equipped with an engine mounted on springs. The car, as I remember, had a torque tube rear axle drive. The power plant unit was pivoted to the frame at the rear end adjacent to the torque tube. The engine front end was bolted fast to the center of a leaf spring that extended outward to the frame members at each side and hung there by means of long spring shackles to the frame members. This allowed the front end of the engine to spring vertically and rock laterally to some extent. While there was a gain in overcoming engine harshness, there were some faults: the engine would continue to bounce noticeably over a bump, but this could be remedied with dampeners. The other was that when the throttle was opened, the front end of the car would develop lateral side shake. Our engineers worked with the inventor to see what could be done to eliminate these difficulties. This work was conducted by a division separate from that researching the rubber mountings.

In the beginning rubber mountings did not seem to hold as much promise as we had hoped. When we neared the time for their release for the coming year's models, we decided we had to expedite things if we were to make any progress. We called a meeting of some 20 engineers. The arguments were hot and heavy. Everyone had his say. The house was pretty well divided as to the merit of the spring versus the rubber mounting. But while the spring type already had been demonstrated in a car, the development of the rubber type had been dragging. I personally argued in favor of the rubber mounting. Finally, the discussion led up to the point that if everyone would concentrate on one approach we could at least hope to have a new engine mount in time for production. To settle matters, all but three of us cast a vote. The group count was still split. So Fred decided to settle the argument. He issued an ultimatum: "Let's all concentrate on the car, and stop work on rubber engine mountings."

As we stepped out of the conference room, Ken Lee came to me and said, "What do you think? Had we better stop working on rubber?" Lee knew of my interest in advocating free action, rubber engine supports.

I came back at him without hesitation, "Ken, you go right along the lines we discussed, but don't say anything to anyone else about it. Keep it quiet."

Ten days later Mrs. Breer and I were headed for two weeks in California. As I left my office, Ken said, "I want to show you something."

I was already late on leaving, so I replied. "Ken, I haven't time to go to the laboratory." Ken continued, "Look, right downstairs by the door is the car I want you to see. Get in behind the wheel, start the car."

So I did. The car was quiet. I said, "What have you got here?"

"Engine on free rubber mountings," he said.

Sure enough he had. I kicked the throttle a couple of times and said, "Ken, you've got a weight shift similar to that of the engine mounted on springs in the other car. Open the hood, and let's see what you have done. Get in the seat and kick the throttle."

Sure enough when the engine was accelerated it kicked laterally or sideways, causing the whole car to shake. This was what we wanted to avoid. I examined the mounting. It was flat, and low at the engine front.

"Look, Ken," I said, "I'll tell you what you do while I am away. Find as near the center of mass of the power plant as possible. Then draw a straight line from the center of the universal joint connecting the propeller shaft through the engine center of mass forward. Where it comes out over the front cross member, mount a cradle-type rubber mounting. Then mount the two rear side supports on a circle using the center line axis as the circle center. Locate the two side mountings at the tangential angle so that engine can rotate freely around the axis line drawn through the center of mass. If you do this you should overcome the side shift action!"

On our trip westward we were resting easy, forgetting the rush I had left behind. When we got off the train on our return, Ken was at the station with the car. Smiling from ear to ear, he shouted, "We got it!" Sure enough, the engine was smoother than ever. The annoying side kick had simply disappeared.

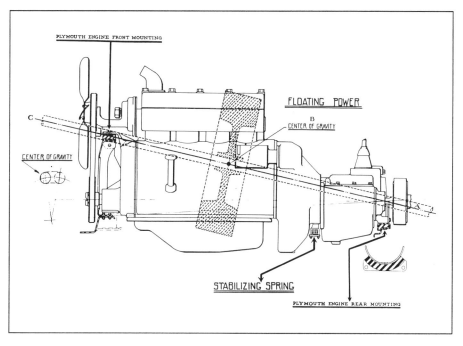

Floating Power was considered a real breakthrough in providing smoother, quieter engine operation, the development of which Carl Breer was justifiably proud. Shown here is a side view drawing of the rubber mountings that made up Floating Power on a Plymouth engine. (Breer Collection)

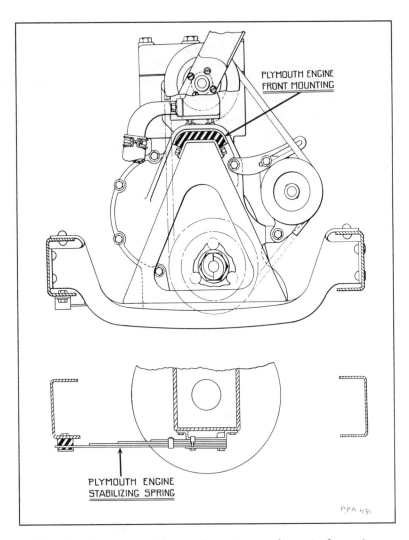

Floating Power on a Plymouth engine, as shown in front view.
(Breer Collection)

Once Fred Zeder found out, he couldn't wait to give it a name. He laid awake nights and one morning came in and said, "Floating power." It was another new innovation. Skelt and the boys all pitched in to its design and testing. The frame had to be changed to accept more torsional rigidity. It was the usual question of an economical balance between cost and lightness, between box section side frame members and the mid-section "X" cross members. Sometimes both were required for maximum rigidity especially on convertible cars. (Ed. note: Floating power first appeared on 1932 Chrysler models.)

The Floating Power front rubber engine mounting as shown in its reality on a bench mounted engine. (Breer Collection)

Floating Power worked out so well for the six-cylinder Chrysler that we applied it to the Good Maxwell Four which preceded the Chrysler Four. (Ed. note: Mr. Breer may be confusing Floating Power with the initial use of rubber engine mounts on the 1925 Chrysler.) With a four-cylinder engine the variation in the explosion impulse forces are far greater than that of a six cylinder. As a result, we had to put a lever arm on the side of the engine, the end of which was mounted in a pair of up and down springs attached to the car frame in order to absorb the excessive torsional rotation of the engine without applying shock to the car. This

Key to the success of Floating Power in terms of the quiet ride it provided was the innovation of a rubber bond so strong that the rubber mounting itself would tear before it would come loose. This publicity shot dramatically indicates how the bond was so powerful that an entire Plymouth roadster could be suspended from a section that had been applied to a simple steel plate. (Breer Collection)

worked so well that we could demonstrate the smoothness of the car operation with one, even two cylinders, shorted out, and there would be very little annoying vibration.

Today we still read where some company has discovered something new in engine mountings, but I would say that everyone has followed our basic innovation. I would be surprised if there were any car engines today mounted on something other than rubber.

Floating Power Applied to Spring Shackles

From our work in the laboratory we had hoped to commercialize rubber, spring-end mountings to replace the lubricated all-metal spring shackles used to provide the movement of the free end of a semi-elliptic spring, an adaptation from the old carriage days.

In a simple sense, all-metal shackles consist of a pair of flat metal links, one on each side of the free spring and eye, and a pivotal bolt and bearing between both links pivoting within mounts attached to the frame. The designs varied: In some cases the links took car load in compression, but in others, in tension. The all-metal link structures called for the making, handling, and the assembly of many parts and much trouble. They also called for constant attention and lubrication to ward off squeaks and excessive wear.

Several approaches were suggested to us by outside suppliers to resolve this problem. The first was developed cooperatively and put on our high price cars. It was known as the rubber kick shackle. It consisted of a special malleable housing that was riveted to the frame. The housing was split horizontally so that the spring ends could be clamped within special molded rubber blocks. On the main leaf ends of the semi-elliptic springs were riveted special shaped stampings. The front end of the rear spring that took the thrust load was shaped so that all spring action pivoted from the front end only. The rear end was allowed free fore and aft movement. Both ends of the springs were entirely encased in rubber, no metal-to-metal contact whatsoever. This type of mounting later was discarded because it was too expensive.

Another alternative type put forward by a supplier was a rubber development known as silent block bushings. It was a rather simple idea, and much cheaper than the one I previously described. It consisted of inner and outer steel sleeves, the symmetrical space between being filled with rubber under high radial pressure. In this unique design, a molded bushing of rubber was used. It was much shorter than the bushings, and its walls were much thicker than the space allowed between the two metal sleeves. A quick action assembly die was used to shoot the rubber between the two bushings. The elastic tension caused by the longitudinal elongation of the rubber when in place was such that extreme high rubber radial pressure was continually maintained between the two steel bushings. The pressurized cushion held them as a unit, yet allowed sufficient torsional movement for spring action to take place between the inner and outer sleeves. The assembly was then pressed into the spring eye, and the two side shackle plates solidly clamped to the longer center bushing by a through bolt. This type was used successfully on the line of our cars. (Ed. note: The rubber spring shackle appeared on 1925 Chrysler models.)

Both rubber types eventually were superseded by an ingenious, simple, economical, all-metal shackle called a "U" shackle developed by Wyc Leighton of the Pressed Metal Products of Port Huron. (Ed. note: Introduced in 1933.) This shackle was made of a well-designed U-shaped forging. The two straight ends carried well machined, coarse threads on which the threaded shackle metal bushings carried the load. One end bushing rode in the spring end and the other in the frame bracket. The V-threads in this case were lubricated by packed grease, and took care of all side thrust, carrying the vertical load and providing free rotating movement without noise or free play. This construction retained lubrication well, and was used over a period of years.

Our Fred Slack ingeniously adapted this screw thread bearing design to our coil spring, front independent wheel suspension where it was used successfully over a period of years, just another first for Chrysler development. (Ed. note: First use was on the 1934 Chrysler Six CA.)

Incidentally, the Fairchild Airplane Company did send us one of their six-cylinder, air-cooled aviation engines so that we could develop rubber engine mounts for a radial-type engine.

The Automatic Choke

Most people today have forgotten that we once had a pull knob on the dash marked "choke." In order to start the car, especially when the engine was cold, one had to pull out the choke knob to force enough gasoline into each cylinder to produce a firing mixture that would start the engine.

The choke valve was arranged so that one could reduce almost all the air supply. In extremely cold weather, one had to be alert to respond the moment the engine fired, just slightly opening the choke valve, otherwise the engine would starve for want of oxygen and flood because the gasoline mixture was too rich for further cranking. The owner then would call for service, but by the time the service man reached the scene, the rich mixture would have disappeared and all he had to do was get in the car, step on the starter, and the engine would start before the owner's very eyes on the first couple of turns.

We decided to do something about it. On our Chrysler payroll we had the father and son carburetor engineers, F. A. and Tom Ball, set up a research program in our cold room where engines and cars could be started and run at temperatures 30 or more degrees below zero. A series of tests were made to see just how much air and gasoline was required to start at various cold temperatures. With this information, a regular starting mixture curve was plotted, first to start, then to quickly follow through on the warm-up without misfiring or stalling the engine.

After much experimentation we then came up with quite a simple answer. First, the butterfly choke valve was allowed to close completely. The choke blade then was provided with a small hole of a size that would allow just the right amount of air to pass through to help vaporize and produce the first firing charge. We gave the blade another, larger air hole with a spring pressured lid to keep it closed. The first, second, and third explosion at the first cranking immediately made the engine step up in speed, forcing the springed air valve to open automatically and fill the right amount of air through for the following firing mixture. These two additions made cold starting exceptionally easy under a wide range of temperatures. Of course, the driver still had to follow through by gradually opening the choke.

We still were not satisfied because it was not foolproof. The owner too often would not push the choke clear back. The resulting rich mixture would cause damage by cutting cylinder wall lubrication and depositing heavy end gasoline in the crankcase, thinning the lubricating oil. Obviously the answer was a fully automatic choke.

The Sisson Company said that they would be glad to work with us, and eventually came up with the answer by gradually moving the choke valve to wide open position via a bi-metal thermostat control. The result proved far better and more dependable than man himself could do. Later, we applied controlled exhaust heat to the mixture during the initial running to help shorten the warm-up period for normal operating condition.

The first winter we provided Chrysler cars with automatic chokes, it was necessary to send a number of men to follow up in the field. Today, all makes of cars and carburetor manufacturers have cashed in on our pioneering work in carburetion. (Ed. note: The vacuum controlled automatic choke initially was optional on the 1932 CP models.)

Development of Rubber Steering Wheels

The evolutionary change from rubber to plastics has been notable in the development of steering wheels. When we first began engineering work on Chalmers production cars, we learned that a company nearby was manufacturing a neat appearing, low priced steering wheel for them. It was made up of a four-spoke spider hub of malleable iron, and a rim made of a mixture of ground up rubber from old tires with what seemed like straw for reinforcement plus a vulcanizing binder. The material was put through rollers to form a rope, then cut to length. The rubber rope was placed in circular fashion around the lock shaped ends of a hub spider in a forming die, then put into an oven and baked. Color mixtures were added to bring out a rather attractive steering wheel. This was an economic gain over that of a strong, laminated wood wheel rim machined to shape and fastened to the ends of the similar spider hub by screws. But it had some problems.

After going through various other steering wheel manufacturing plants to see how they were made, we decided to check the rubber rims for strength. We found that if you leaned on them too hard, they would collapse and break. Certainly this would not do. We complained to the makers but were rebuffed. So we established safety specifications for rim strength that they would have to meet. They could not, and went out of business. They were replaced by the Lobdel Manufacturing Company, a north Michigan firm, who had developed an attractive, composition wood rim wheel many times stronger than the original Chalmers unit.

We felt that sufficient rim strength provided safety by cushioning the driver and not breaking away from the hub in case of accident.

The Dryden Rubber Company of Chicago through their Detroit representative, Magnus Burgess, worked very closely with us on a new rim, and formed a large business making attractive steering wheels of most any color, molded to any shape. The rubber rim gradually evolved into the type using flexible plastics.

Development of Fatigue Analysis

As a result of the excellent showing that Chrysler's early stock cars made in the European LeMans races, Europeans' attention was called to our high grade car, one that sold at a low price but competed favorably in endurance and speed with the most costly cars on the Continent. However, we ran into difficulty in Antwerp. There were several accidents. In one case, there was a lot of publicity because royalty was involved. About that time Walter Chrysler was headed for Europe. He said to Fred, "Better come along. We'll go to Antwerp and see what we can find out about their wild driving."

When they arrived, they discovered something different: it was not the fast driving. To the contrary, it was the continuous pounding that Chryslers were taking over the unusual Belgium cobblestone roads. Instead of being flat like our early split granite roads, they were quite rough because the cobblestones were large and round.

Samples of steering knuckle arms were quickly shuttled back to us. All were broken at the same critical point of the arm lever to which the drag link connected to the steering gear. Naturally if steering was lost, the best the driver could do was jam on the brakes, and hope for the best.

It was a serious situation for us, but there were no other roads on which we could test like those in Antwerp except in Chicago where some of the streets also were paved with imported Belgium cobblestone. We decided to build our own simulated cobblestone road in the form of a testing machine which we named the "Belgian Rolls." We set up two heavy steel shafts, wheelbase distance apart, and mounted on each of these two large, three-foot diameter, wide-faced steel pulleys car wheel tread distance apart with a third crowned belt pulley amidships between the pulleys on which the car wheels rode. The car to be tested rode on the four pulleys. The four corner cables and adjustment turn buckles kept the car in position while it ran under its own power. The car wheels drove the rear shaft while the amidships heavy leather belt would drive the front shaft at the same speed, just as would a car running on the road.

To simulate the pounding of the cobblestones, we simply bolted two or three, two-by-four inch hard wood blocks across the wide pulley face parallel to the driving shaft. As soon as the driver started the car and the pulleys went to work, we knew we had acquired the desired test action. Soon we replaced the driver, and produced the up and down accelerator action by automatic mechanical means. Lo and behold, in less than 24 hours of steady running, we produced the breakage in the steering knuckle arms just as did the Belgium cobblestones. In fact, the test condition was so severe that we would even blow out several tires before a steering arm would break.

As a result of this simulation exercise, we started two additional laboratory divisions that are of major importance today — the stress and strain laboratory, and the fatigue testing laboratory.

The stress laboratory used polarized light on transparent plastic shapes to show up stress lines. Stress lines are like contour lines that wind around on a surveyor's map showing the variation in steepness among the hills and valleys. Constant stress lines are sharply shown in the same manner. These lines move in and out as stress is increased or lowered. By making a replica of the steering knuckle arm out of clear plastic we could see where there was a concentration of forces leading to breakage. The steering arm had a stop shoulder lug extending outward at one side as an integrally shaped part of the forging. The stop shoulder intentionally limited the car's turning radius to keep the front tire from scuffing against the frame. This extending lug had a sharp corner where it extruded outward from the forging. Metallurgical etching showed clearly how the steel grain in the forging swept sharply around this corner. The stresses were concentrated and stacked up like the map contour around the steep corner of a mountain peak. We call this a notch effect. For example, if we meant to break a piece of wood at a given point, we simply take a knife and notch it at that point. Concentrated binding stresses will cause it to fatigue quickly and break at the weakened point. So it was with the steering arm. Only in this case, the grain made a sharp turn in the corner of the protruding lug adding to the concentrated weakness when alternating under straining forces. We modified the forging, and many successful checking tests were run on the Belgian Rolls. These Belgian Rolls are still an important piece of our laboratory equipment. Modern variations of this simple laboratory setup can be seen in all competitive plants.

The important results of research and laboratory work in the automotive field cannot be underestimated. An accident caused by the failure of an axle shaft, wheel spindle, or steering mechanism seldom, if ever, happens today. Also, tire blowouts are rapidly fading into the past.

Evolution of Tires, Wheels, and Rims

Owen Skelton followed the development of tires, wheels, and rims very closely. He was responsible in great part for gradually bringing the overall diameter of the wheel down, and the diameter of the tire cross section up with a resultant lowering of air pressure to cushion the road. In the early days, tire rims were split, with an outer flanged ring that was bolted by lugs to the wheel rim holding the tire in place. Then we used a snap lock ring to hold the flange in place of the bolts and lugs.

A big advance later was the drop center, one-piece rim in which part of the tire casing wall could be dropped into the center, thereby providing clearance for the rest of the wall to slide over the larger rim flanges. The inside air pressure would then hold the tire casing flanges tight against the rim flanges and inner shoulders. It was an ideal situation except for one hazard: In the case of a blowout, the casing walls would shift off the supporting shoulders and drop into the center well. Then, under road pressure, the tire casing would come partly off the outer rim flanges, invariably jump crosswise, and become all tangled up with the wheel. It would revolve in a "humpety-hump manner," causing the car to strive to get out of control.

I have never forgotten the experience Mrs. Breer and I had in this regard. We were driving up a long stretch of straight, paved highway following a canal in Florida on the approach to Jacksonville. There was not a car in sight. We were in one of our large, eight-cylinder Airflows. We had to make a railroad schedule, and were a little behind time. Also, I was curious to see what this particular car would do at high speed. So I let her out. We were rolling along about 90 mph when the right rear tire blew. Using the brakes as carefully as I could, I slowed the car down to 55 mph when the centrifugal force of the tire quit supporting the car. All of a sudden, the tire dropped flat, and jumped the rim. It took all of my strength to keep the car from veering and plunging into the deep canal alongside of the road. When we stopped, we saw that the tire was wrapped up over the rim, and on both sides of the wheel. The inner tube was torn into shreds, and had created a severe dragging force at the right rear.

Shortly after this happened Skelton and I visited at the Goodyear plant in Akron, Ohio, to take an inspection trip through the mammoth Zeppelin lighter-than-air ship. After enjoying breakfast at Mr. Litchfield's home, Litchfield, who headed Goodyear, took us to his office for a chat. On a sample table were various short-length cross sections of tires and steel rims. One in particular attracted our attention. When we asked what this was, Mr. Litchfield said, "That is a section of the Dunlop tire used on a rubber-tired experimental railroad train in France. They are trying to interest us in making them."

The unique aspect of this design was that in case of puncture or blowout, the tire would flatten or drop only a half inch or so before it would be riding solid on a heavy raised section of aluminum on the inside. The object, of course, was safety, and the necessity of keeping the train on schedule to reach the next station. I immediately spoke up and related in detail the close call I had with the Airflow, and, what's more, I could visualize the development of a similar safety feature on motorcars which would especially benefit high price cars.

Not very long after our visit, Goodyear put on the market the "Life Guard" safety tube. It consisted of a double inner tube, the inner one smaller than the outer tire diameter with fabric walls sufficient to hold full air pressure. The all-rubber outer tube, having a common wall at the rim section, expanded outward eccentrically, applying full air pressure to the outer tire casing. Air was pumped into the smaller inner tube which, by means of a small fixed orifice, allowed air to flow into the outer tube. In case of a blowout, the casing and rubber tube could split open, but the strong inner tube would hold air pressure long enough to support the car until it came safely to a stop. (Ed. note: Chrysler adopted the Life Guard tires in 1936.)

Today, with the development of synthetic fabrics that are much stronger than nature's own cotton, especially at higher temperatures, blowouts may soon be considered a thing of the past. Add to this the rubberized inner layer of the tire casings, manufactured always to be in compression, and the puncture hazard has been removed. These modern casings press onto the leak-proof steel rims tight enough to retain air pressure better than former tires with their costly, air retaining inner tubes. This rapid advance of stronger tires that run cooler and have longer life is a marked improvement.

During our time automobile wheels have evolved from large diameter, all wood spoke and wood rims, into wood spoke and steel rims, then into an all steel wheel pattern. Also, early tires and rims were interchangeable units when removed from the wheel rim. Now they have evolved into a simple, removable disc steel wheel unit with tire, the brake drum, and hub being retained as a fixed part of the car. Wheel evolution had really been ahead of the tire development until now.

Improving Brakes

A big advance in brake drums was made by a unique method developed by the Campbell-Wyant & Cannon Foundries and Motor Wheel company of Lansing, Michigan. A steel drum with a centrifugally cast iron brake drum surface was bonded to the steel as cast. This furnished a cast iron wearing surface with the backing strength of steel. It resulted in a much greater life without surface scoring. There were other semi-steel or so-called alloy brake drums, satisfactory but usually heavier. (Ed. note: Centrifuse drums appeared initially on the 1932 Plymouth.) At the same time, brake linings of reinforced woven asbestos brass wire gave way to a molded to shape type with asbestos fiber in the mix. The first brake linings were riveted to bands or shoes. Now through our development efforts with a cement or cycleweld bond, the lining has become an integral part of the brake shoe, eliminating rivets.

The Change to All-Steel Bodies

Our first closed bodies were made by the Widman Body Manufacturing Company, formerly a showcase manufacturer. Bodies in those early days were made of wood encased in steel paneling, whether for a closed or an open car. This was a natural outgrowth of the horse-drawn carriage business.

The metal outer coverage of the bodies were nailed (screw-fastened where necessary) to the hardwood under-structure, pillars, etc. Overlapping seams were covered with trim molding. Metal parts were soldered for strength or soldered for finish fill. Metal brackets and braces were screwed fast at the junction points for strength and stiffening. Also, hardwoods were getting to be more scarce. The Briggs Company, like most others, had bought mammoth stands of timberland and set up saw mills, all to protect against future wood shortage for body building. Our corporation had done the same, as had G.M.'s Fisher Body, and Henry Ford in northern Michigan.

When Widman finally quit manufacturing auto bodies, we took on the Briggs Body Corporation to make our bodies. Our contract with Briggs was for them to furnish bodies on a fixed price basis. The price evidently was low because when the closed bodies were delivered for volume production, we discovered they had a structural weakness, a so-called weaving and squeaking.

First we called in Herman Maize, the Briggs Body Superintendent, and showed him the kind of stuff he was shipping us; the glued wood joints had not been given time to set. Many of these joints in the new finished bodies were so loose we could easily make the body weave. We told Briggs that if they could not do better, we'd look for a new source. It wasn't long before a new superintendent, Henry Hund, came over, as Briggs probably saw the "handwriting on the wall." To our surprise their attitude had changed overnight to a "Let's get together and see what we can do about this one." We must have added over a dozen, various shaped, malleable brackets all over the body and added bolt fastenings to the wood screws to furnish maximum rigidity. We were making progress, but regardless of how well we would build them, we were continually fighting squeaks and rattles both in production and when the car was in the hands of the owner.

To escape this dilemma we began to pressure our body suppliers for an all out steel structure. Budd Manufacturing Company had pioneered bodies along this line in England. Briggs Manufacturing finally began to realize the importance of cooperating with us, and before long, we had steel bodies in production.

Not long after, Fred Zeder and I had to make a trip to New York. We decided to see how our largest New York dealer, Colt-Stewart, was getting along. Although Stewart happily welcomed us, he was rather cool about our new line of steel body cars. (Ed. note: In all probability these were 1933 Chrysler Royals.) Stewart said, "You know, people come in to look at the cars. While they like their style, they claim that they are weak because the doors are tinny sounding when slammed shut." Stewart added, "We can't sell them. What's more, in case of damage, they think they will be expensive to repair."

At that time there were several luxury carriage builders in New York who were famous for the special luxury car bodies they put on motor car chassis. These bodies, like the closed type of horse drawn vehicles, invariably sounded solid when their heavy doors were shut.

Fred and I argued, "Jim," (Stewart) we said, "there's no comparison in strength. Our bodies are lot stronger and safer."

Yet Jim said, "We know all that, but you can't argue against the doors sounding tinny. Come on up and we will show you." He was right. We slammed car door after car door. They seemed lighter, and when slammed, the door and car panels would resonate.

"Jim," we said, "we have to catch the Detroiter (train). Why don't you come with us to Detroit? Your personal appearance will force our boys to realize that we need to take quick action."

Jim came with us, and the next morning we held an early meeting with our body division headed by Oliver Clark and his assistant Tommy Thomas. Out of this meeting we instituted a new sound deadening laboratory to develop new manufacturing sources of sound deadening material. The tinny door panel sound was deadened in the same manner that a torsional vibration crankshaft dampener was able to do its work. Like a boy in a swing, an opposing retarding force can keep him from swinging. So it was with a door. A cheap, weighty material bonded to the inner wall of the door panels did two things: It added weight simulating that of the old coach door; and when the door was slammed, it had a solid sound like the coach door. Today, sound deadener is scientifically applied where needed throughout the body. It includes undercoating which when added to the body floor pan and fenders adds quietness and stops road resonance as well.

These kinds of laboratory efforts helped put our all-steel bodies over the top, and are the foundation on which all automobile companies are built today. No production car today, outside of a styling sample, is ever built of a wood/steel composite construction. This applies to motor trucks and trailers as well. The custom, luxury wood body is a thing of the past.

Competition was slow to follow. For example, Fisher brothers, a part of General Motors, had set up a most attractive demonstration of car body building at the 1933 Century of Progress Exposition in Chicago. It was meant to sell the public on the fact that composite, wood steel bodies were the way bodies should be built. It was illustrative of fine Fisher Body craftsmanship, notwithstanding the fact that the General Motors Fisher Body Division had investments in lumber reserves and fine, woodworking equipment. The fact of the matter is it is impossible to bond steel to wood so that the relative movement between the two can satisfactorily be eliminated. The rubbing or chapping between the two was a problem too difficult to handle. The annoying squeaks and crunches were difficult to eliminate. Wood is far better suited for the construction of homes and such where the forces of relative movement are negligible, and where undamaging wind noises may be construed as music in the realms of protection.

Our body structures laboratory grew rapidly. Today it is a very important factor in designing a suitable structure of maximum strength. The end result of all this is that it is now far better and cheaper to make bodies of steel. Today I would not know where to find a piece of wood in the modern automobile, and squeaks are passe.

Once our all-steel body practice was established, our next step was to find the most satisfactory, most economical way to rustproof it. We accomplished this by applying a chemical dipping process developed by the Parker Rust Proofing Company of Detroit. Today, all sheet metal in its final body-in-white form is dip processed to assure a long life against rust. (Ed. note: Chrysler considers this a 1931 "first.")

Valve Seat Inserts for Long Life

We were alerted to the need for a new type of valve seat insert material from experience with a truck line hauling castings from Ohio Foundries to our Detroit plants on practically a 24-hour basis. These were heavy-duty Dodge trucks of the type used for hauling an overload capacity with trailers. The engines would operate practically wide open without stop hour after hour. The contractor was happy with the operation, but the grueling effort produced an epidemic of cast iron cylinder blocks in which the exhaust valves had pounded down their seats. We went all out to remedy his difficulty regardless of cost by incorporating Stellite valve seat inserts. What we did was machine out the already worn valve seats in the cylinder block to accurate size, then let the Stellite rings that had been prechilled below zero to expand tightly into place. The remedy was so successful that our metallurgists followed up with various types of cast alloys until we found a much more economical metal insert. In fact, we ultimately reached a point whereby we could release exhaust valve inserts as standard equipment in all our passenger car engines, another advanced step toward long-life engines. (Ed. note: The special steel alloy exhaust valve seat inserts appeared on 1933 models.)

Powder Metal Research Leads to "Oilite"

In our early engine design, a cylindrical sleeve bushing of bronze was pressed into the flywheel end of the engine crankshaft into which the extended end of the transmission shaft carrying the disc clutch rested. When the clutch was released, the bearing bushing would revolve about the shaft holding everything in alignment. Excess lubricating oil from the engine unfortunately would cause clutch slippage, as would packed grease, graphite, etc. The location being inaccessible, there was no simple way to force grease in from the outside. Normally it was only a matter of time until the bearing would become dry, and growl and squeak when releasing the clutch. Oilless bushings were suggested as a solution. G.M.'s Moraine Products was making a bushing of powder copper and graphite compressed into a solid bushing at extremely high physical pressure. We tried them, but many crumbled when being pressed into place. They were too sensitive for such application and there did not appear to be any way to strengthen the structure.

Mr. Sherwood stopped by one day with an idea he thought might have merit. Sherwood's background was in the development of the then well known "Cutless Bushing," molded rubber, water lubricated propeller shaft bearings for which he held the patents and was receiving royalties. The "Cutless bushing" still is a popular all-rubber underwater propeller shaft bearing used in motor boats. Grooves in the bushings allow water to lubricate them, and, surprisingly, they last many years.

Mr. Sherwood's proposal was for a feasible new way to make a powdered metal bearing. If it worked, it would be far superior to G.M.'s Moraine bushing. We hired Sherwood and put him to work with one of our best research metallurgists, Bill Caulkins. After a few months time, he and Caulkins developed a bearing of powdered copper and powdered tin compressed in a die to the desired shape under high pressure, then heat treated in a furnace in a non-oxidizing atmosphere just sufficient for the tin to melt and make a strong bond between the copper particles. The result was a structure astonishingly high in physical strength compared to former bushings which one could almost crush by hand. Moreover, the new bushings had about a 40% porosity by volume, such a porosity that one could blow smoke straight through the metal. The unique part was that the porous areas could be filled completely with oil by subjecting the bushing to a high vacuum, submerging it in oil, then allowing the atmospheric pressure to force oil into the pores. The pores were so fine as to be hardly visible without a magnifying glass, yet they were continuous, and afforded a large surface area to retain oil. These new bushings, wiped dry, would show no oil on their surface, yet temperature or friction would bring the oil to the surface.

This development not only was the answer to our problem, but the real beginning of our "Oilite" powdered metal parts division which brought in many millions of dollars of profits to the Chrysler Corporation. (Ed. note: Oilite bearings began to appear on the 1932 models.)

When Mr. Chrysler was next in Detroit, we appealed to him for a sales outlet for the many things being developed by the engineering department, so that we could have an outside market for such products as powdered metal parts.

It was not long after that that the Amplex Corporation was set up for this purpose, with the "Oilite Division" being the first under the new arrangement. Our work led to a variety of miscellaneous metal powders, both of ferrous and non-ferrous materials, such as bushings for bearings of all shapes and sizes of iron and copper powders for gears, oil pump rotors, door lock parts, etc.

One day I said, "Let's try to make a porous powder strip metal in continuous form to replace the expensive woven fine wire mesh used as gasoline strainers." The suggestion I made was to build a machine with two, large size, polished rollers, both power driven with an upper hopper by means of which the mixed powder metal could be fed between the two rolls, and squeezed out in a continuous thin sheet. After several months, the lab boys came in with samples of thin powder sheet but their porosity was not uniform enough to be used as a strainer. However, it proved to be of great value when applied to continuous sheet steel strips which, after heat treating, could be formed into steel back type bearings.

But Caulkins and his associates were not licked. They came back later and showed me samples, both in sheet form and various shaped types, of a means of making globular powder of uniform small size. Having accomplished this, it was only a matter of sintering these small spheres into flat or other desired shapes. The strainer porosity was of uniform size, and more effective as a gasoline strainer than the expensive wire woven type. The effectiveness of the filter was proved dramatically by one of our executives who ran out of gasoline in the country one day. He stopped at a nearby farm house, and told the farmer that he had run out of gas and asked for a small pail of water. The farmer graciously complied. As a matter of curiosity, he walked out to the car, and saw the executive pour water into the gasoline tank. To the farmer's astonishment, our executive stepped on the starter and away he went! What really happened was that our new strainer was so effective that it had let the gasoline through but not the water. When the water was added into the gasoline tank, it raised the little gasoline left in the tank to where the strainer would let it through, enough to let him make it to the nearest gas station.

This powder metallurgy research project started in my office in the summer of 1927. To date, 1965, Chrysler Amplex Division has maintained its leadership as the largest of the some 150 such manufacturers in the world.

Chrysler now turns out more than 6,000 different powder metal products, many of which go into the power train that Chrysler warrants for five years or 50,000 miles.

Helical Gear Transmissions

Our car transmissions were a subject on which Owen Skelton and Harry Woolson concentrated much effort. The three speed and reverse straight spur gear type was standard, and most popular with the public for a long period of time. The gearshift pattern likewise was standard from the very start in the automotive field. The refining progress was evolutionary beginning with a better application of roller bearings, then with synchronizers that would bring the gears up to meshing speeds quickly just previous to engagement. Design ingenuity successfully made it easy for everyone to shift gears quickly without clashing.

The next improvement was the adoption of skew gears making transmissions entirely free from gear noise due to the rolling action of the teeth as the driving torque was smoothly passed in an overlapping manner from one tooth to another. In addition, the skew gear teeth were structurally stronger than the straight spur type. The shifting of skew gears was made possible by the ingenious use of a skew splined gear shaft having the same spline twist as the gear teeth. With the skew gear design, gear shifting was as easy as with the straight spur and straight spline transmission. This was another great first for Chrysler that competition soon adopted. (Ed. note: Chrysler introduced helical gears on its 1933 CT and CO models.)

At this time, the European trend toward multiple speed car transmissions was not acceptable in the U.S.A. Only trucks would submit to, or rather welcome by necessity, four or more speeds in which to shift.

One year we developed a four-speed transmission that had a regular three-speed and reverse shift pattern, then by releasing a latch button on top of the gearshift, the lever would push further forward to engage a step-up gear above that of direct drive. It was an innovation we introduced on our top priced car for speed and economy. The public would not take to it so we dropped it the following year.

Another invention was one in which the clutch released the gear train at each end, allowing one to mesh any pair of gears easily and quickly without the pressure clash of engaging gear teeth noise. It was a very quick acting and fast engaging four-speed transmission.

Another school featured a feathering spline pickup type. Upon release of the clutch, a feathering key could be revolved from the drive shaft to engage any gear set desired, quietly and quickly.

Still another was the planetary type where gears were always running in mesh, and all one had to do was to friction band engage the gear ratio that he desired to use. This type was popular on European makes of English and German Daimler cars, but there were friction losses from all gears running at different idle speeds, and the added complication of band pickup at that time. Our engineering division also designed and built an exploratory one of similar kind.

While this experimentation was going on, the three-speed and reverse standard shift pattern of synchromesh transmission was used on all makes of cars for a considerable period of years.

Keller Overdrive Development

Despite our unsuccessful past experience in attempting to interest the public in a fourth gear, we became attracted to an idea patented by a Mr. Keller, an inventor on the West Coast. Keller offered the possibility of making the fourth speed independent of the three-speed gearshift pattern.

The idea was mechanically simple in principle. We called it the Keller clutch. It consisted of a sliding pawl set to move radially outward from the rotating transmission axis. A coil spring held the pawl inward against a stop. At a predetermined rotating speed the centrifugal force of the unbalanced weight of the pawl would overcome the spring, and the pawl would quickly slide outward to engage a slot in the cylindrical surface of a member of the fourth speed planetary gear set, thereby locking out the fourth gear ratio. Normally the three-speed spur helical gears would be driven through the reduced fourth speed of the planetary gear set. We named the planetary gear set "overdrive." When these planetary gears were locked out the car would be running in direct drive, meaning no gears were operating, and the engine would be turning over in the least number of turns per car mile. The name "overdrive" is misleading in a technical sense, but to the public it meant another drive or change in gear ratio over and above that of the three-speed transmission.

Normally, the spring tension would be set to give way at some speed, say 38 mph, above which the pawl would spring out and attempt to engage the slot on the gear set. If the slot would be moving by too fast, the pawl would skip over until it had a chance to engage. The driver could choose at any speed above 38 mph to get into overdrive. All he had to do was to release the clutch by pushing down on the clutch pedal with his left foot, then removing right foot pressure from the accelerator pedal. The clutch speed then would slow down to where the pawl could click into the engaging notch, and the car would then be in overdrive with the clutch re-engaged. In reality the car now would be traveling faster than when engaged in the former direct drive for the same engine speed. The pawl would automatically release at speeds below 38 mph if the driver let up on the foot throttle. It was a great achievement, and we first used it on the Airflow automobile. It made the Airflow glide along quietly at high speed with seemingly little effort or noise. The contrast in driving improvement caused the inexperienced driver sometimes to catch up too quickly to the car ahead because he did not realize the smooth high speed at which he was driving. It was a complaint that we received when we first introduced the Airflow car from some drivers who found themselves into the backs of others. We had to issue caution warnings to the new owners.

One Saturday morning, I was trying to clean up my desk to get away by afternoon when Karl Pfeiffer, one of our top experimental engineers, came in with a long-drawn face. He stared at me for a moment, then blurted out, "Mr. Breer, this overdrive is a mechanical monstrosity. All it does is pound itself to pieces. When the pawl jumps across the slot, it pounds the corner away like a hammer just before it engages. It's like putting a piece of steel in a vise then hitting it continually with a hammer. The metal won't stand up."

Just about that time it dawned on me that we had a parallel experience with valve tappets. Early valve cams were designed to allow the valves and tappets to pound down on their seats to stop. We cured the fault by first stopping the valve and tappet short of its seat, then just letting them fall the rest of the few thousandths of an inch.

I told him, "Karl, do the same with this as we did with valve and tappets. Bring the pawl to a radial stop, then as the engaging slot comes by, give the oncoming landing surface ten or fifteen thousandth outward radial clearance more or less. This will give the lowered surface a chance to get under the pawl end before centrifugal force can move it that far, thereby eliminating any hammering of the engaging slot edge."

This proved to be the answer.

The real accomplishment was Pfeiffer's finding of the design weakness which could have cost a lot of money if not caught before production.

In the operation of research and laboratory work, I found it more important to lead men rather than to direct them. If the individual showed a lack of progress, it was much better to reassign him to some other project better suited to his liking.

This was the basis of our sound engineering approach to creating harmony during our regime of successful automotive pioneering for our Chrysler Corporation.

The importance of resolving the Keller clutch's problems could not be overestimated. It was a necessary component of the new Airflow. Let me explain why. The Airflow requiring full body structural strength in place of a rigid frame was new to our body design department. As released for production, the body was much heavier than the sample road cars we tested. Why? Because each one of the designers added a little metal thickness "just to play it safe." Our body draftsmen at that time did things that were mostly intuitive since they were not educated in structural analyses. The added weight was a drag on our engines. They did not have sufficient power to give the desired low- as well as high-speed performance unless the Keller clutch or overdrive assisted them.

A serious question then arose as to whether we should manufacture the clutch or give it to the Borg Warner Corporation. This was an issue for Walter Chrysler to decide. On our next trip to New York, K. T. Keller, who headed manufacturing, and I presented our case to Mr. Chrysler. My decision was that we simply had to have the overdrive with the Keller clutch.

"How much will it cost for us to tool up for the clutch?" Chrysler asked K.T.

"Twenty-five thousand," was the quick reply.

Without any debate Mr. Chrysler said, "We can't afford it; let Borg Warner make it."

From then on Borg Warner worked out a deal with our corporation that put them in the overdrive business in a big way, but it also opened up our patented overdrive clutch to use by our competition. Our competitors quickly adopted and featured the Borg Warner Overdrive. To the trade and public, it was known as the Borg Warner Overdrive.

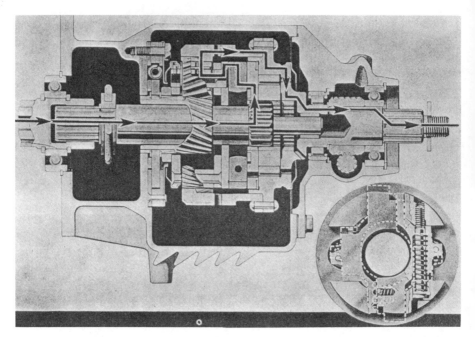

Another Breer breakthrough was in taking the Keller clutch and adapting it to Chrysler cars as a fourth gear called "overdrive." Overdrive was introduced on the Airflow cars and allowed them to glide quietly along with seemingly little engine effort. Chrysler then released the overdrive patent to Borg Warner rather than tool up for an in-house build. (Breer Collection)

Development of Fluid Drive

Our next pioneering advance was the fluid drive. It involved a unique, two-vane, stamped metal impeller, one for the drive, the other driven in series with a pedal-operated clutch connecting the engine. At low engine or idling speed, the vane impellers had negligible torque or driving ability, but as the engine revved upward, the driving torque increased rapidly, allowing the driver to stop and start a car without releasing the clutch. In fact, one could start while still in high gear. All he had to do was step on the throttle, and the car would accelerate smoothly. The slip at the low end allowed the engine to run up to a speed immediately where the engine torque was greater. In that manner, a quicker getaway was accomplished. At first, there appeared little interest outside of engineering in the device until one day Tobe Couture, Fred Zeder, and Skelton took Walter Chrysler on a demonstration run to show him how the car handled around town where there were many stops and starts. When Mr. Chrysler came back to Engineering his first question was, "When can we get that in production? Just think, you can drive around town without using your clutch." We named it "Fluid Drive," another first for Chrysler. (Ed. note: Fluid Drive debuted on the 1939 Custom Imperial.)

Experimental Studies in Automatic Gear Shifting

One day Ken Lee and one of his engineers came to my office to settle a heated argument as to which ideas for automatic gear shifting I would prefer to have them design and build. Ken said he wanted the car to shift automatically from low to second at a given car speed, and from second to high at another fixed car speed. The argument was that one of his engineers felt strongly that it would be much better to have the car shift from low to second after so many feet of travel, then shift to high at another given distance.

I liked their enthusiasm, so I said, "Look, I can't settle your argument, but I think I can suggest an idea that would give you the answer with little expense. Build two automatic transmission devices, try them out, then come back and tell me which is the best. Take a standard car, equip it with a speedometer and a distance traveled meter on the back of the front seat, put in a clutch pedal, a gearshift lever and accelerator in the back seat compartment, and try out variations of both your ideas. Put a third man in the back seat, give him the pattern, and have him reproduce both your ideas of an automatic gearshift."

A few days later, Ken Lee and his associate came in my office with smiles on their faces. From the results of their tests they decided they were both wrong. However, this by no means limited our desire to accelerate a car as quickly and smoothly as would a steam car.

The problem with the manual, friction clutch type of shifting was that it was too irregular and jerky when the driver transferred power from one gear to another.

After World War II, car transmission development turned toward a combination of multi-vane hydraulic turbines operating in series with friction band pickups of planetary gear ratios. They utilized self-contained oil pumps driven both from the engine and rear axle drive shaft which in turn would furnish governor-controlled hydraulic pressure to automatically apply the various gear ratios.

We soon had a two-speed, fully automatic transmission of this type developed and practically ready to release, but were a little hesitant about the heavy expenditure needed for production equipment at the time. We held it back for a few years until competitive pressure finally forced the issue, and a simplified, two-speed, selective type was developed and released to production.

Engine Exhaust Muffling Innovations

Originally the muffling system was a back pressure type which, through gradual resisting expansion chambers, reduced pressure, cancelling out the resultant sound. It called for 25 to 30 psi back pressure, meaning that the engine had to push against two times atmospheric pressure to exhaust the already used gases into the air to meet the requirement of minimum exhaust noise.

It worked so well that muffler cutouts became popular on the accessory market. The reduced back pressure added to the performance and top speed of the car. The loud bursting explosion noise to the atmosphere added to the thrill of having more power and speed. Some states soon passed laws forbidding the use of cutouts, especially in the city. In spite of this, more cutouts were sold.

While the public was putting pressure on the law enforcers to stop this public nuisance, I had a hunch, called in our muffler engineer, and asked him to fit up one of our mufflers with a valve at the engine end of the muffler that would allow engine exhaust to go straight through the muffler center chamber and out the tailpipe as well as through the outer muffling expansion chambers to the tail pipe. I felt certain that this would eliminate the objectionable bark people complained about as well as relieve most of the back pressure.

Tests were made of my concept. The results were as I had hoped. The muffler was quiet enough for it to be accepted in the city, yet there was no loud shrill bark in wide open country driving. What's more, back pressure was reduced to a negligible amount.

The next muffler development became known as the Burgess type. It was a muffler that you could look straight through from the engine end to the tailpipe without restriction, meaning practically no back pressure. The pipe wall through the center of the muffler had a lot of evenly spaced small holes punctured through it. The cylindrical space between the center tube and the larger diameter outer muffler casing was filled with steel wool. As a sound deadening principle it was simple and very effective. The explosion peak wave pressures of the engine exhaust would pass out through the holes into the resisting steel wool space, then pass back in to fill the low pressure area of the same wave immediately following. This dampening effect was so complete that the exhaust gas flowed out into the atmosphere smoothly without noise. This type of muffler was very popular until the present, longer life type was developed which was no more than substituting a multiple number of resonant chambers of fair size that hold shape and did not clog. Had we as engineers been alert at the time, today's muffler would have evolved without the Burgess muffler, but mufflers then were of secondary importance, and were mainly engineered and supplied by outside manufacturers.

Automobile Wheel and Tire Developments

Our O. R. Skelton has led the way from the original large wheel, small section, high pressure tires to the development of the small wheel, large section, low pressure tires of today. Originally we used demountable, snap ring rim 30" wheels and six ply 5.77" tires with inner tubes in 1924 that operated under 35 lb pressure. The wheel sizes were gradually reduced step by step. In 1931 we adopted the simple drop center rim feature, thus eliminating the snap ring means of holding the tire casing in place. The tires at that time were the four-ply 5.77" size.

Today wheel size has been reduced to 14 in and equipped with larger size 8.25 two-ply tires with 24 lb pressure.

Skelton's alertness resulted in him bringing back from the west coast a patented development which our laboratories evaluated and released on all our cars as an important contribution to safety. This was known as the hump rim. A cross-section of the drop center steel rim shows a raised hump on the shoulder edges adjacent to the rim flanges. The tire casing walls snap tightly over the humps when the air pressure is put in. In case of a blowout or a sudden flat tire, the tire casing walls remain in place and do not drop into the drop center rim groove keeping the casing from turning crosswise which could jam car motion. (Ed. note: Safety rim wheels were introduced by Chrysler on its 1940 models.)

When we pioneered the running balancing of wheels and tires, we ran into opposition with tire manufacturers until they were sold on the importance of eliminating the careless overlapping of fabric layers in tire making.

As a side light, our Research Department headed by Ken Lee projected the concept of a tubeless tire that would be vulcanized or bonded to the steel rim. Years later, a member of our tire supplier company stated they did not know why we did not follow through at the time because tubeless tires are universally used today.

Development of the All-Glass Sealed Beam Headlights

In the design of the Airflow car, our objective was to set the headlights flush in the car body, a problem because of lens size.

At the time we seriously considered using the "Woodlite" headlamp developed on the California coast. This lamp had an oval egg-shaped body with a small aperture lens opening through which the light beam projected. First we had to have the approval of the New England State Supervisors Engineering Committee's requirement of light distribution as set up by the State Highway Commissioners Group. This novel design lamp could not meet their specifications, so we had to use what then was standard. Typical headlights at the time, as I remember, had a lens minimum diameter limit of 7-1/2". We compromised by using two separate single beam lamps one above the other. This allowed us to have a one inch smaller or 6-1/2" upper lamp for country driving and a smaller lamp below for city driving.

Since then I had been on the lookout for headlamps of a smaller size that we could adapt.

One day I told C. Harold Wills about my desire for the development of smaller headlights because of the experience we had with the Airflow. Harold Wills, formerly with Ford, had given up the manufacture of a car he had designed, the Wills-St. Clair, and joined our Chrysler Corporation.

Harold said, "You know, I am well acquainted with General Electric's Nela Park Lamp Laboratory in Cleveland, and there is a research man down there I know very well. His name is Enfield, and if you like, I will have him come up here to find out what can be done about smaller headlights."

One afternoon, Mr. Enfield, Harold Wills, Ken Lee, and I sat down to review the various reasons why headlamps had to be of such large diameter. Mostly it was because large size parabolic reflectors were necessary to compensate for the accumulation of manufacturing limits that had to be allowed in making the various parts in a headlamp assembly design of that time.

While Mr. Enfield described the problems of variable limits, I made a sketch suggesting the possibility of using a longer life bulb, presetting the filament for the main driving headlight beam in assembly, then soldering the socket to the reflector so that the filament would glow in the correct focal center of the parabola.

I signed the sketch, dated it, and passed it over to Ken Lee to sign and said, "Hold this for the record." Our habit had been to do this with new ideas for patent record purposes. I noticed at the time that this had bothered Enfield.

In closing our session I said to Enfield, "Why don't you go back to Nela park and just forget that you ever saw an automobile headlamp? Start from scratch. See what you can develop by projecting light through the smallest lens possible."

Some time later Harold Wills stopped in my office and said, "Just came back from Nela Park and saw Enfield. Asked him whether he made any progress with the headlights. He said, yes he did. He had made a lot of progress. He had showed what he had accomplished to his boss, Mike Sloan, the director head of Nela Park, who told him not to show it to a soul, not to anybody. Then he told Enfield that his accomplishment was the greatest advance in lighting in years, not only for automobiles, but for indoor and outdoor distributed lighting."

What Enfield did was to mount the two light filaments as a molded part of the glass reflector on its own glass pedestals such that they would be held or could if necessary be adjusted to the exact focal center of the parabola. The filament was coated with aluminum. By subjecting the open reflector to a high vacuum in a fixture, and the heating filament to incandescence, the aluminum would vaporize, depositing a mirror like coating on the perfectly smooth parabolic glass reflector walls. The lens would be made of die molded glass with many prism surfaces on the inside, refracting the light beam to properly meet the driving and passing road light distribution according to the state specifications.

The lens then would be placed against the parabola face in a rotating fixture where the respective edges would be melted into unity by means of highly concentrated heat flames as the lamp revolved so that it would be removed as a sealed beam unit. The lamps then would be filled with an inert gas so as to retain maximum brilliancy over the period of the life of the lamp. Previously light bulbs had to be removed quite often. With the sealed beam, filament life was intended to be three times that of the replaceable bulb of the past, making the service cost of light renewal approximately the same as before.

The all-metal auto lamp manufacturers could see the handwriting on the wall as an all-glass lamp was much cheaper to make and more efficient because light passed through only one glass wall instead of two. This new glass lamp idea created excitement. It was like a bombshell to the industry because its merits and low costs were so outstanding.

Now the battle was on to make sealed beams the universal standard of headlighting for the entire industry, and to convince the association of State Highway Commissioners to set up appropriate standards.

The Hall Lamp Company, being a branch of General Motors, took the sealed beam secret to G.M. headquarters which caused O.E. Hunt of the G.M. Corporation to call a meeting of the automobile manufacturers. As a result, a committee of engineers was set up for the first time under the auspices of the A.M.A. (Automobile Manufacturers Association) to handle such various engineering matters that pertain to common interest of all manufacturers of automobiles.

The Henry Ford organization at that time operated as an independent. Although not a member of the A.M.A., nevertheless they had their engineering representative sit in on the committee meetings and took their share of expense.

This became the start of an alliance of engineering known as the Liaison Committee which consisted of the top engineering members of the automobile manufacturing companies.

Committee members were O. E. Hunt (acting chairman), and his brother John Hunt, both of General Motors Corporation; Col. Jesse Vincent of Packard; Northrup Bates of Hudson; Harold Youngren, who sat in for Ford; and myself, who represented Chrysler Corporation. Also included were B. B. Bachman from the Auto Car Corporation while Sherman represented A.M.A.

As an industry, several millions of dollars must have been spent in development. Many tests had to be run at night involving many people, including the organization of Highway Commissioners. General Motors was very cooperative in lending their proving ground men and facilities whenever needed.

In coming to a common lighting standard the question of the number of light beams had to be settled. G.M. at the time was using a three-beam system(one for the country highway, one for urban use, and an in-city beam. Chrysler standardized on a two-beam system known as country and city driving beams. The A.M.A. committee, after running many tests individually and collectively, adopted the two-beam system. It was rather fortunate that the present sealed beam headlight system has been standardized, and we soon forgot the varied lighting qualities we had to contend with previously such as when the headlamps aged and the reflectors blackened or the necessity of replacing short life light bulbs. Europe was rather reluctant and slow about accepting American sealed beams on account of the many European specialty lamp makers that added an individual style to their lamps to make them attractive.

Further development evolved into the present lighter and simpler unit construction lamp with similar self-contained parabolic reflector providing very efficient directional light spread.

In brief, our foresight led to the development of the sealed beam headlight system, and no doubt added no small amount of income to General Electric Corporation. (Ed. note: The sealed beam headlamp began appearing on passenger cars of the 1940 model year.)

Development of Amola Steel

The name of C. Harold Wills brings us to Amola Steel, a development of Wills and our group of metallurgists headed by McCleary. Harold, through his early development of vanadium steels while with Henry Ford, had established valuable contacts with the heads of various steel corporations, while McCleary had an unusual ability in various types of steel analyses. Amola Steel was the result of both their efforts, a comparatively low cost steel, highly competitive with the various expensive alloy steels on the market at that time. Harold was the man who went out and aroused the steel company's interest in Amola Steel and its manufacture. The steel was classified as an abnormal steel. It was of unusually high quality and had a very fine grain. When Harold expounded on the fineness of Amola grain quality, I remarked, "It should be fine for razor blades." As a result, our Oilite powdered metal division used Amola Steel, Gillette-type, double edged razor blades as an Oilite advertising medium. (Ed. note: Amola steel was featured in the Airflows.)

The automobile industries in reality have been at the forefront in the actual development of all types of steels as well as many high quality alloy steels to meet the various unusual requirements on an economic basis.

The volume of steel consumed by the automotive industry is rather tremendous when one realizes that approximately 70% of the weight of the average automobile is derived from steel, with around 10-11% of the car weight consisting of expensive alloy steel.

Introduction of Our Fresh Air Heater

One of our developments adding to passenger comfort was the fresh air heater. As a research project, it was assigned to Allen Staley.

Previous air warmers were developed as an accessory, and were of the re-heating type; that is, they consisted of a fan that would recirculate the air in the car through a hot water radiator. It was much more popular than the previous exhaust gas type heater which invariably added an undesirable, burned, dusty smell through the car. On the other hand, the hot water type recirculated the humid air that came from a person's breath which, in cold weather, precipitated troublesome, inside window frost.

The concept of a fresh air heater was a "natural" for motor car use inasmuch as cold humid air, when heated, becomes drier. The colder the outside air, the drier it becomes when heated. The solution was simple in concept — develop a way for outside air to blow through a hot water heater, distribute it properly, and control it automatically once adjusted by the driver. For defrosting the windshield, all that was necessary was to apply the hot dry air against the inner surface.

The best location for an aperture to take in fresh air was above the cowl, hence the development of the cowl ventilator for incoming fresh air.

Our first hurdle was to develop a rain trap so that air could come through separately from the water which was to be drained to the outside. Before we worked out a satisfactory method to do this, one of our laboratory heads came in the office and said he did not think it could be accomplished. He was wrong.

Our initial fresh air heater was simple in principal but became complicated when the sales department demanded that we design it as accessory equipment so that dealers could install it later if desired. As a result, the heater was considerably penalized for we had to design an attachable rain trap beneath the standard cowl ventilator, add another door for summer fresh air, equip it with two blowers pulling the air through two separate hot water radiators, then deliver the air at high velocity rearward through air chutes neatly closed in at the cowl sides. Also, hot dry air had to be delivered to the inner surface of the windshield for defrosting purposes when necessary as well.

The fresh air heater system, while costly, turned out to work very satisfactorily. Two air chutes delivered the heated air past the front seat ends very effectively, heating the rear compartment. Also, one could keep one's feet comfortable by moving closer or away from the heated air stream. Our limousines also delivered heated air through ducts in the front doors to rear glass partitioned compartment. (Ed. note: The air trap appeared on 1939 models.)

As the public became more heater conscious, demand for the fresh air heater use grew to the extent that now factory built-in systems are far less complicated and controlled by automatic thermostatic temperature controls.

Competition followed, first by putting the heater equipment under the hood, and bringing fresh air from the front of the radiator fairly close to the ground so as to save the cost of a cowl top inlet. Today everyone agrees that the logical place for an air inlet is on top of the cowl.

One year our younger group of engineers following heater developments decided it would be more economical to follow competition by taking the air intake forward of the radiator. They insisted that they had test driven our vehicle with this system around the city, and that the air was as good as that coming from the cowl inlet.

But while driving on a Colorado street with my heater fan running one season, a fellow with an old car stalled in front of me, and could not start his engine without being pushed. So I kindly accommodated him. The moment his engine fired my car was filled with his vicious, rich, smog-making exhaust gas. When he drove on, I had to stop and get out of the car for fresh air.

When back in Detroit, I told the boys that I had a new name for their heater, "Let's call it the 'Fowl Air' Heater." They came back with, "Next year's model we are moving the air intake back to the cowl ventilator where it belongs."

This is a typical example of what can happen when you follow competition too blindly. It was the engineering policy of Zeder, Skelton, and Breer to be pioneers, not followers. Before too long, all of our competition also came to the conclusion they were wrong and began to move their fresh air intakes on top of the cowl ahead of the windshield.

Kettering's Ethyl Gas

One day Walter Chrysler came in and stated that Ket (Kettering) had tried to interest him in adopting ethyl gasoline. Chrysler said, "I wish you would take a look at it and see if it is all that he claims." Ket had been trying to interest the various General Motors car divisions in the virtues of ethyl gas, but, as in the past, they were indifferent to his "anti-knock" claims, so he turned to Chrysler.

As I remember, we took Tobe Couture, Jack Macauley, who later joined the Ethyl Gas Corporation, and a couple of mechanics with us and equipped two sedans identically — one of which later would be equipped with a higher compression head plus an added 300 lb of weight for ethyl gas tests. From previous engine tests and calculations we knew that if ethyl gas would work as claimed, a high compression car could carry an added 300 lb, the equivalent of two extra passengers, with the same acceleration and hill climbing ability as a car using conventional gas less the 300 lb. To avoid any misunderstanding we had tetra-ethyl lead added to the same basic gas that would be used in the lower compression engine.

The anti-knock performance was astounding. Up hills or accelerating on the level, the two cars ran neck and neck. We came back to Detroit with enthusiasm. Use of ethyl was a much better solution than what we had done in the past when we had released engines with more costly cylinder heads made of aluminum just to get the increased performance of higher compression.

As a result, we took a high compression, cast iron cylinder head, enameled it bright red, and advertised the phenomenal performance results of our "Red Head" engine. The Red Head went over big, and gave Ket and his ethyl gas a great boost. (Ed. note: The Red Head became an option on the 1928 Imperial 80 and 80L.)

In retrospect, the Ethyl Gas Corporation maintained headquarters in the Chrysler Building in New York for many years in appreciation of Chrysler's initial interest.

Electric Auto-Lite Corporation Becomes a Dependable Spark Plug Source

When our major electric equipment supplier was taken over and absorbed by our major competitor, we could not find a satisfactory supplier to replace him that could furnish us the automobile electric equipment that would meet our requirements for quality and dependability.

Finally the Electric Auto-Lite, a rather rundown company that prospered when Willys-Overland was in its prime, was sponsored by some of our directors as a possible supplier providing we would put our ability and guidance behind the company. P. J. Kent and McKechnie headed our electrical department, so their organization was directed to step in and do this job.

Eventually Auto-Lite reached the level of a satisfactory supplier, and began to expand and prosper. Now Auto-Lite rates as one of the Big Three spark plug manufacturers.

In our early association with automobile developments, spark plug problems abounded. Spark plug success always centered around the most satisfactory insulation of the central electrode. For many earlier years a battle was waged between porcelain insulation or an assembled center body of mica-compressed discs. Today mica spark plugs are few and far between, if not entirely eliminated. Porcelain has taken over.

We noted several times how Royce Martin, heading Auto-Lite, said he would like to get into the spark plug business. It just so happened that a ceramic engineer by the name of Robert Twells came in to see us one day looking for a job. After reviewing his background I called Royce Martin on the phone and told him, "If you want to get into spark plug manufacturing, I have the man you want in my office right now!"

The next day Royce hired Twells. Every year I get a long personal New Year's greeting letter in appreciation from Twells along with a summary of the great progress he has made.

Electric Clocks and Alternating Current Generators

Some years back we initiated projects to develop a more accurate and more dependable electric clock, and to adopt simple alternating current generators to replace our direct current generators with their inherent commutator and carbon brushes difficulties.

We never quite succeeded with the clocks. It appears progress will lag until efficient electronic devices are developed at reasonable cost.

Alternating current generators, on the other hand, were bound to be adopted as soon as new avenues are opened up for less costly rectifiers. This just recently has been accomplished. This year (1960), Chrysler set up its own factory, and alternating current generators with efficient rectifiers have become standard equipment on all Chrysler Corporation cars. These generators should eliminate past electrical service problems. They have a great advantage over the D.C. generators by providing a charging current supply at engine idling speed, an important advantage in city driving.

Chrysler and We Three Engineers Enjoy a Luncheon Visit with Henry Ford at the Dearborn Plant

When our Plymouth line of cars was introduced, Walter Chrysler planned a trip to visit Henry Ford to show him our new Plymouth car.

Mr. Chrysler had the three of us engineers (Zeder, Skelton, and Breer) as a party of four call on Mr. Ford at his engineering headquarters in Dearborn early before noon. Mr. Ford gave us a most cordial welcome. He personally showed us around the building where the engineering divisions were arranged in open area fashion. At one extreme end were a few dynamometers, at the other was an enclosed room where he proudly rejuvenated the old fashion dances — waltzes, the one step, mazurkas, square dances, etc. with old time musicians. Mr. Ford was very proud of this area and the pleasure of doing this sort of thing.

When it was time for lunch, Mr. Ford led us through this spacious building over to their famous round table Rotunda dining room. There we met with Edsel, Charles Sorenson, Martin, and some others. The spacious table seated some 10 or 12 people.

The conversation mainly centered around the industry. Mr. Chrysler told them about our Plymouth line and suggested they examine the car we brought with us.

During the drift about cars and engine testing, I rather enthusiastically spoke about how we had set up laboratory testing and developed our engines on the dynamometer stands where we could exaggerate the tests to such extreme condition as 50-hour tests. I also dwelled on how we flow tested for carburetor development and were running our economy research road tests on the dynamometer where we could control all the conditions that would effect the accuracy of test results.

In reply Mr. Ford said, "While we have a few dynamometers, we do our testing on the roads; that's where they have to run anyway." Naturally I remarked that we also carry on extensive road testing in addition to our indoor development testing. Seemingly Mr. Ford's interest was not in dynamometer work.

Mr. Chrysler was anxious to show Mr. Ford and his men our Floating Power feature on our four-cylinder Plymouth that gave it both a quiet and smooth operation. We learned later that Mr. Ford could not develop any interest in our engine rubber mounting feature, yet today it would be difficult to find any production car not equipped with the floating power type of engine mounting.

At the end of the dinner, fair-sized portions of pie were served. After taking a couple forkfuls, Fred Zeder spoke up, "Mr. Ford, what does this pie consist of?"

Mr. Ford, who was a vegetarian, said, "That's carrot pie, Fred. Do you like it?" And before Fred could respond, he added, "Walter, bring Mr. Zeder another piece of carrot pie!"

Later when anyone asked what Ford served for dinner, the answer came quickly, "Carrot pie." It became a byword. Whenever Fred would get enthusiastic about some objective, someone would interrupt with, "Fred, how about some carrot pie?"

After dinner, Mr. Ford took us to a little secluded laboratory building. It was an electrical laboratory devoted to the development of electrical equipment for automobile use. When entering, you saw 10 or 12 bench setups with middle aged men operating little high speed steam engines which in turn were driving electric auto generators or ignition equipment. The plant had its own boiler setup as well as its own electric lighting plant. In other words, it was a completely independent, fully equipped unit entirely free from outside dependence.

From here we were taken through Thomas Edison's laboratory which had been moved and restored in every detail from Menlo Park, New Jersey. Ford enthusiastically showed us around the various stages that led to the development of the electric lamp. Then when asked about Edison's reputation of working without sleep, Mr. Ford showed us where Thomas Edison, when tired, would lay on the floor with his head and half his body in the closet under the stairway so as not to be disturbed from a well-earned nap.

When we came to the heating boiler, Mr. Ford opened the furnace door and said, "You see that fire? I had Edison start that when he was here, and I have arranged to have that fire burn forever!"

Mr. Ford had a lot of sentiment about human beings and their accomplishments as verified by the realistic collection of buildings at Greenfield Village as well as the restored historic equipment of the museum.

One could not be more gracious than Henry Ford in his all out effort to make the day most interesting for our party.

Chrysler Institute of Engineering Evolves

Our Chrysler Engineering division had been growing so rapidly that bringing in outside engineers just seemed to increase the load on the three of us rather than to relieve the burden. In 1928, we had over 500 engineers under our supervision. At that time we were proportionately larger than our competitors in engineering personnel.

The real problem was that when we brought in an experienced man from competition it would take a year to get him down to our way of doing things. With inexperienced engineers, it was a time-consuming educational problem.

The cure we thought would be to set up a student apprentice system, a graduate apprentice feeder of engineers from the various universities to our corporations.

My further thought was that we should find the ideal man to handle and direct the men through their apprenticeship. It would be a full-time job working with the apprentices.

We spoke to Mr. Chrysler who suggested we talk to K. T. Keller (president at the time). This we did and K. T. said, "Bring the candidate over to my office and we will set it up."

Our problem was that we did not want the program set up as a feeder for the whole corporation initially since we would have the unions to contend with as a barrier. This would be especially true in our assembly areas.

So we set it up to start as a feeder to Engineering, and later would let it expand to the production end. We had plenty of various separate divisions that could schedule men through a two-year training course, give them lots of diversification, and at the same time handle any problems that were bound to come up.

While this was being put together, we sent inquiries out among our men for an ideal leader to head the setup. After a few months' search, we found a man with the desired qualifications working in our Dodge plant. He was John Caton, who had been the professor directing the automotive engineering department of the University of Detroit, and had taken a leave of absence from teaching to gain experience in our plants. We found him in overalls with a large wrench in his hand doing inspection work on the Dodge Truck line. After he agreed to our proposal, we sent him to General Electric, Allis-Chalmers, and Westinghouse to acquire up-to-date experience in apprenticeship courses.

Some 20 engineering graduates were approved by Mr. Caton and brought in for the first time. We assigned them to work in the various laboratory divisions.

After a few days, the fellow in charge of our mechanical laboratory came in to see me and asked who these two college men were who don't even know how to handle a wrench. He had had to move two of his mechanics to another division to make room for the two, and put them to work on the bench. He was astonished when we told him to bring back his two mechanics, and that the two new engineers were added to his division for training.

Unfortunately this momentary displacement of two mechanics for two technical men had aroused much gossip and conversation throughout our engineering organization, and gave the impression that we were going to replace the mechanics and others doing drafting work with those who had college degrees. It took years to live this down.

As time went on, the college apprentices began teaching the practical men the things they wanted and needed, and the practical men in turn were teaching the college men the things they learned in the school of experience.

Our drafting rooms were filled with men who came up the hard way. Many of them started from the bottom as blueprint boys. Calculations were done primarily in long hand. Slide rules and logarithm tables were not considered tools of good craftsmanship. The college apprentices changed this.

John Caton then arranged a Michigan State charter to award both high school and college degrees. An educational board and organization was set up for operation under the name of Chrysler Institute of Engineering. The Institute grew and expanded into building quarters adjacent to Engineering. Later, under Harry Woolson, the corporation constructed a properly designed school building with class rooms and physical, well equipped chemical laboratories. The building was located on Oakland Avenue adjacent to Central Engineering and had its main entrance on Oakland Avenue with proper insignia over the door.

Breer had a special affection for the Chrysler Institute of Engineering which first opened its doors in 1930 thanks to his far-sighted prompting. He felt that the Institute would short cut the time necessary to train engineers and draftsmen to the Chrysler way to doing things. (Breer Collection)

Graduation day for the Chrysler Institute was one attended by highest management as shown here. From left to right: Fred Zeder, Walter Chrysler, Carl Breer, Owen Skelton, and Byron Foy. (Breer Collection)

Several hundred university graduates selected by our engineering sponsors were brought in for the two-year course.

This Chrysler Institute started in 1930, and many of its graduates went on to hold top positions in the corporation. While starting as an engineering feeder of men, it evolved into men taking positions in all branches of production. It is now a self-perpetuating organization in that the Operating Board of Directors consists mainly of Institute graduates. Alumni members personally go to the various universities to select desirable graduates, then take the responsibility of directing and guiding these men during this course of two years training.

The importance of the two-year feeding of this Chrysler bloodstream of men into the corporation after graduation is that they are better equipped to take on added responsibility in the large central engineering organization. When they graduate, they are fully acquainted with the many engineering departments and their respective responsibilities. Should they move into the car manufacturing divisions, this experience of two years in engineering is of great help in whatever operation they pursue.

Reminiscences of Early Product Development at Chrysler Corporation   141

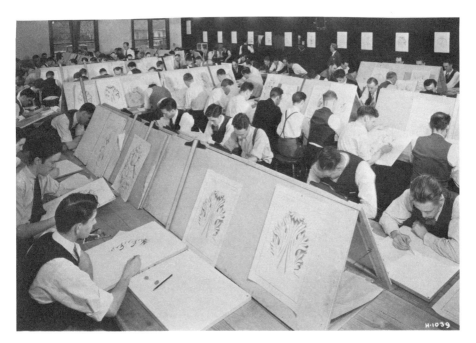

Shown here are two of the classes in session at the Institute,
the Body Drafting class (above) and the Chemistry Class (below).
The Institute was to provide Chrysler with numerous graduates that went on
to high management positions over the years. (Breer Collection)

The introduction of graduate college engineers in some of our engineering divisions took more time than in others, and initially caused us some concern. One reason was that the draftsmen resented technical men coming in for drafting experience. On the other hand, technical men welcomed those with laboratory experience.

In the Chrysler Institute drafting and art instruction classes were set up hoping to encourage drafting vocation. Today these men are in many top corporate positions and have become more and more in demand by the various divisions outside of engineering.

The influx of these men lessened the overload on the three of us and at the same time strengthened the sound centralized engineering foundation invaluable for the corporation's future.

Part V:
Birth of the Airflow Car

Exploring the Role of Airflow Patterns

In the late Twenties, Mrs. Breer and I regularly spent the summers on the shore of Lake Huron where our boys could bathe and play on the sandy seashore at Gratiot Beach, Port Huron. This was about 60 miles away from our engineering offices at Highland Park. I would spend the weekends at the beach. If the days were hot during the week, I would often drive up after work and be back in the morning. The road most of the way was straight. Gratiot Avenue was only partly paved and the rest was macadamized.

One late summer Saturday afternoon at dusk, I was rolling out Gratiot when I noticed a flock of geese flying fairly low across the highway at some distance. The sight intrigued me for the moment because it was too early in the season for geese to be flying southward. On further observation, I realized that what I was seeing was a group of Republic Army planes flying low in formation from behind a group of trees.

This got me to thinking. The planes seemed to stay in the air at such a low speed. I got to wondering about airflow over and under a car — whether the air crowding under the front of the car would lift the front end up so that I would have less steering hold on the ground, or would the pressure be downward and overload the tires. I put my arm out of the window and was surprised at the amount of lift that my forearm would receive as I rolled along. Yet if I angled my forearm further upward and outward from the elbow, the amount of retarding or back pressure that resulted became noticeable. One could almost feel the car slow down or pull to the side. When I got back Monday morning, I called in Bill Earnshaw, an engineer of means from Dayton, Ohio, who had just joined us. Bill was fascinated with research.

I told Bill of my experience Saturday and suggested that he take one of our cars and extend some indicator wires between the body and the front axle and explore if they created air lift power or downward air pressure by simply changing the normal height space between the front axle and the car frame. The next week Bill came in and reported that there was no lifting or downward effect that he could measure, but he did feel a head-on wind resistance that was quite appreciable. Bill, whose home was in Dayton, personally knew the Wright brothers, and had seen their air flow work with oil and lamp black. He suggested that he go to Dayton and talk to the Wright brothers who had accomplished much experimentally at little expense by using a blower.

I set him up temporarily at home in Dayton, and in a few weeks he came back with some metal plates simulating an automobile against which were fastened various angular board shapes. When a lamp black mixture was released, it would deposit air streaks on the metal plates showing how the air would sail up and away as it passed over a square corner, and what a difference a rounded corner would make. It also showed how the air would twist and swirl in eddy currents at the back of the automobile, as it tried to close in behind the simulation of various car shapes.

We Build a Wind Tunnel

We decided to see what would happen to a car in a wind tunnel. Wind tunnels developed for aeronautical purposes were expensive to build. They could run close to a hundred thousand dollars. Earnshaw went back to the Wright brothers in Dayton and Wilbur Wright gave him a lot of encouragement in building an inexpensive tunnel. In a few months we had a nice, compact, aerodynamic lab located in a 20 x 40 ft space equipped with woodworking equipment to build various shapes of small models. The wind tunnel throat was about 20 x 30 inches in cross section. The wind creating propellers were V-belt driven with a DC variable speed motor capable of about 35 hp.

The first few models were replicas of our present production cars. We also reproduced some of the exaggerated shapes which Bill had demonstrated with lamp black and oil to correlate the indicated eddy currents with relative wind drag. The results were very interesting. We found that we could follow and duplicate various air swirls with a thread on the end of a rod, moving it about while watching through a side window.

Cadwalder and Gail Borden, an aerodynamist, were our laboratory team. They had just completed tests on two production models. While we were going over the data, I remarked, "Let's find out what the drag is if we run them in opposite directions. It could be that the air resistance would be less when they run in a reverse direction."

Birth of the Airflow Car

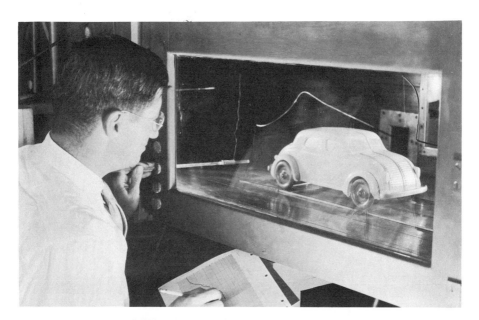

Breer's inquisitiveness about air flow around the automobile resulted in Chrysler building the wind tunnel with a 20 x 30 inch throat shown here during the late 1920s. At first wood replicas of production cars were tested during which Breer and his team discovered that then current models encountered 30% less wind resistance if run in a reverse direction. Gradually they began to modify the vehicle shape, the one shown above having the characteristics of the Airflow pre-prototype design called the "Trifon." (Breer Collection)

Wind tunnel tests with small scale models led eventually to experiments with full size streamlining by extending the rear of a vehicle via a long tail. The developmental version shown here was reported to have been driven at 100.5 mph. At this early stage, the top speed achieved by an Airflow not having such extended aerodynamic structure was 62.5 mph. (Breer Collection)

To our astonishment, air resistance was some 30% less when we did this. Our laboratory was on the fourth floor of engineering. I looked out of the window at all the cars below and remarked, "Just think how dumb we have been. All those cars have been running in the wrong direction!" This astounding enlightenment was the motivating force that started our design and development of the Airflow series.

As a result of this work we decided that the main automobile shape should be like that of a zeppelin or blimp with a large oval section forward that would slowly taper toward a smaller section at the rear. Basically it was a question of having the air close in smoothly behind the traveling object rather than spreading it at the front.

Air has weight. The farther the air is disturbed outward away from the car, the less power is left to the engine to drive the car forward.

Our analysis indicated that to improve a car body for less wind resistance it would be necessary to move the bulk forward from the rear. Our first step was to move the rear seat located over the rear axle to a position ahead of the rear axle where we could lower the rear passenger without diminishing comfort, otherwise, with the standard car design of that day, we would have to move the rear axle behind the rear seat, thereby lengthening the wheelbase.

I called in Oliver Clark, then head of the Body and Styling Division, and told him to lay out a full-size longitudinal section of a car on a chalk board to see how we could develop a more aerodynamic car. We assigned Cadwalder to make the blackboard studies. We started first with the rear wheel location, then put the rear seat ahead of the rear axle. A front seat cross section was sketched in on the same lower level as the rear seat cross section. Then, at a proper leg room distance for the driver, we located the sloping footboard and pedals, then the dash. We placed the front wheels as close as possible to the dash and footboard with minimum wheel clearance. This established our wheelbase, now even shorter than our previously accepted standard.

"Where are you going to put the engine?" was the next question. If we placed the engine in the rear, it would mean a complete tear-up, and the car would be rear end heavy unless we limited engine power to hold down engine weight. A car with a heavy rear end on the road would act like an arrow that was shot with the feather end forward. So it was an engine in front or nothing. Traditionally, front mounted engines had been located behind a cross-beam type front axle, the object being to have it low enough to be in line with the propeller shaft. That did not suit us, but to relocate it forward of the front axle was out of the question. There was only one place left — over the front axle, a first in our industry.

Our next step was to lay in a body profile. With headroom markings for passengers, we drew a parabola encompassing the front end of the vehicle upward over the top making the highest point over the front passengers. For proper vision, we placed the windshield fairly close to the driver which required stepping down the parabola to the hood line but rounding it downward over the engine and front wheels. The center line or horizontal axis of the parabola passed approximately through the wheel centers.

To favor streamlining, the gradual taper from the front to the rear reversed the customary seating order, putting three passengers in the front seat and two in the rear. Chrysler claimed that sales would suffer too much from competition if we did not have a three-passenger seat in the rear, so we compromised by making the sedan a six-passenger car. Since the coupe was acceptable when seating three in front and two in the rear, we did taper the coupe more freely both in profile and in plan view toward the rear.

Car length forward from the passenger compartment and windshield was dependent upon the length of the engine. Thus, the DeSoto with a six-cylinder in-line engine could be shorter than the eight. Of course, "V" eights were still in the future.

At the time the aeronautical world was split between two structural schools — one that favored structural beams and depended less upon the outer metal surface for strength; the other favored a strong outer strength of metal and less on inner reinforcing metal structure. The latter, known as "monocoque" type of structure, was our preference.

The new Airflow body layout, with front axle and wheels closer in, paved our way toward adopting the monocoque design. By eliminating the removable hood, we were able to extend the body sides forward for structural strength, and at the same time picked up a wide windshield which placed the "A" posts outside of the most desired line of vision, improving the driver's view of the road.

Oliver Clark and I decided that a flat windshield of this magnitude would look terrible. A curved windshield with less wind resistance would be stronger than the flat. And if we eliminated the customary center division post which allowed the two-piece flat windshield to angle forward, we would have uninterrupted vision forward. We called Herb Sherts, the Pittsburgh Glass research engineer, and discussed the idea with him. This was a tough assignment because windshields were made of Du-plate safety glass which consisted of two layers of highly finished plate glass with a strong transparent plastic bonded between them, a standard requirement for windshield safety at that time. Sherts made the curved windshields experimentally by allowing the pair of matched plates to sag in pairs through gravitation one on top of the other, to the shape of the fixed surface below when heated uniformly at just the correct temperature. All of our first series of experimental road test cars had curved windshields, and we had to develop multiple blade wipers for them. Both are listed as Chrysler firsts!

Previous to the Airflow, all front engine cars had their front axles ahead of the engine with semi-elliptical springs mounted on the axle. All had heavy car frames that had to project out front to hold the front spring pivot end. To save both weight and cost, this design favored short extensions which meant using shorter leaf springs as well. This resulted in a front end jouncing frequency of around 120 cycles per minute. European cars had a noticeably higher frequency — as high as 135 per minute. The rear end of all cars were sprung at lower frequency.

With the Airflow design, we were able to use longer front springs. As a result, we lowered the natural frequency by about 25% or to around 90 cycles per minute. The ride felt as if you were changing from a running pace to a more comfortable, natural walking gait. It proved to be such a shift in comfort that all competitive car makers eventually changed their spring rates.

In our streamlined Airflow design we used a front lifting, lid-type hood to cover the radiator and provide access to the engine. This did not leave much room to set the headlights flush in the body structure between the fenders and hood lid as was customary for that time. What's more, the minimum lamp diameter that then had been adopted as standard by all states was 7-1/2 in. We managed to get acceptance by using a slightly smaller lamp with a separate passing light.

Front end overhang, too, was unusual for its time. The first all-metal prototype ready for production was displayed in the top floor engineering showroom. It was here that we presented the Airflow to Mr. Chrysler for his approval. Things went fine until we sat down in chairs facing the side of the car. Mr. Chrysler looked at the front end for a few moments, then said, "Too much overhang at the front end." This caused quite a last minute tear up for us which we later thought might have been avoided if only we had sat Mr. Chrysler facing the front end. This overhang criticism being the only suggestion, we were happy to cooperate with Mr. Chrysler's wishes because most of the time he was enthusiastically in agreement with our designs and objectives.

Coordinating Airflow Body Design with Production

Putting a monocoque structure in mass production was a new problem for manufacturing. After much consultation, we decided to use a very light chassis frame so that the cars could be assembled just as they had been in the past, with the frame carrying all assembled parts along the assembly line to where the body was set down on it. The new light frame had practically no torsional or beam structural rigidity. It was then bolted up underneath to the body using rubberized fabric shims. The unit one-piece monocoque body was designed from end to end like a bridge with internal trusses that would be more rigid and solid than any other body/frame construction. To keep the rear end of the car clean and smooth, we made the trunk space back of the rear seat available by tilting the rear seat back forward. We did mount the spare tire under a metal cover on the rear of the car.

The method of building a monocoque body which production developed was also something new at the time. What we did was hold various body stampings in place and automatically weld them into a unit by means of multiple electric spot welders operating in unison. The various unit assemblies then were located in a large fixture where they were welded into a complete body. This process of group spot welding was new. So much development work was required resulting in extensive delays.

The interior ceiling involved a different process, too. Instead of using the customary soft woven cloth lining, we applied a smooth bakelite or hard Formica material that had previously been formed to shape and held fast in place against padded insulation by molding strips.

Another innovation was chrome plated, tubular seat frames. Special trim fabric seat backs were fastened to brackets attached to the tubing leaving the chrome tubing free around the cushion edges in armchair fashion. It was something new and attractive at the time, and called for an entirely new shop department that had to be educated in how to buff and polish the tubing through its many operations to attain the desired chrome plate highly polished finish. The tubing had to be surface finished, given several coats of copper and nickel plating, then polished before applying the chrome, after which it again was polished. This called for much manual work in handling and holding the tubing against buffing wheels. There were not enough experienced grinders and polishers in the whole of the Detroit area to fill these requirements.

At the time we had a very outstanding scientist and engineer, Dr. Carl Huessner, as the head of our Research Electroplating Laboratory. Since the cost of chrome plating the tubular seat structure ran higher than anticipated, Dr. Huessner researched special means for accomplishing a natural polish in the plating process which saved considerable time and effort in the manual labor operation. He introduced these changes in the production alignment and spent personal time in assuring that the plant operation would be successful in production. As a result, he was accused of taking work away from the men on the line. Doc's life was threatened, his family was threatened, and his car and home were both damaged by the heavy missiles. The corporation was helpless to do much about it but did protect him with body guards for several years.

Later, he was recognized for his pioneering work in making it possible to extract fissionable uranium isotope U-235 from U-238 to produce atomic material in quantity. Dr. Huessner was one of about 12 men who received a citation for his work on the atomic bomb.

Developing the First Airflow Prototype

Our very first experimental Airflow car was built without either a clay or wood "mock-up." It was designed more from an exploratory viewpoint to assess the rearrangement of shifting both the front and rear axles back some 20 inches from the generally accepted standard of that time.

Our first car carried a six-cylinder engine on a relatively short wheelbase. Its development and build were kept very secret. The design as well as the construction were done in isolated quarters so as not to interfere in any manner with vehicles being readied for production. This work had been going on for several years with no intent to consider it for production until we knew more about what the new weight distribution would produce.

The earliest experimental design version of the Airflow was the Trifon, built in the late 1920s. The riding qualities of the Trifon were considered to be so outstanding that they were adopted for the much more streamlined Airflow to follow.
(Breer Collection)

The Trifon shape was a product of the Chrysler wind tunnel, but its implementation into a full-size working pilot model for the Airflow was the responsibility of Oliver Clark, shown here discussing its merits with Breer. (Breer Collection)

Since we dared not drive the vehicle around Detroit, we worked out a deal with a farmer, of the Strubble family, who had large acreage in upper Michigan on a back road off a secondary highway some miles north of Grayling. This family had previously furnished sleeping quarters and meals for a few people during trout fishing season on the Au Sable River, a branch of which ran through their farm. We arranged to set up a small garage and workshop behind the house, and conveyed the new car to it from our Highland Park plant under cover of darkness in a large enclosed trailer. It was a unique setup. Many testing miles were run on the surrounding country highways with various later model cars without even attracting any curiosity or attention. At that time this was isolated country.

Birth of the Airflow Car

On the night before our first ride, Fred, Skelt, Tobe Couture, A. G. Herreshoff, and a few others went up to stay at Strubbles'. The van with the car arrived early the next morning. Ramp planks were set up, and a new species of car was introduced in all secrecy to be seen for the first time in the northern wilds of Michigan.

The four of us got in and went for a drive with other cars following in case of trouble. What a surprise! The road seemed to be smoother than when we came north earlier. And the forward view was wide open. The front pillars that always appeared to be in our way now were out to the sides. Riding in the back seat seemed just as comfortable as in the front. Most astonishing was the relaxed feeling that came from the ride. One could put his head back against the cushion and relax in perfect comfort. All the usual bouncing, shaking, and pitching had disappeared to be replaced by a slow, relaxed, gliding movement. Independent wheel suspension on American cars had not as yet been developed commercially, but we had the same sensation one would get from such a suspension. It was a great day!

Late in 1932 and during the summer and fall of 1933, road tests of Airflow prototypes were held in two secret locations (actually the Strubble and Mead farms) near Grayling, Michigan. The Airflow prototype was delivered by van as shown here, usually at night, then taken out upon the back roads for grueling tests far from prying eyes. (Breer Collection)

Imagine the shock the local farmers must have experienced in 1933 to see a vehicle with the radically different shape of the Airflow pass them by. The two automobiles shown here look as if they belong to two different eras, yet in a sense they were contemporary. In all, three prototypes were built for testing, two with six-cylinder engines, and one with an eight. (Breer Collection)

We continued road testing all that spring and into the fall. We contacted Mr. Chrysler in New York and asked him if he could spare enough time when next in Detroit to spend the day with us at Grayling. We did not go into detail because we wanted to get his personal reaction to the Airflow prototype, whether it would build up the same enthusiasm for its ride to him as it did to us when riding over the same back roads.

As planned, Chrysler came. After having a nice, home cooked dinner at Strubbles', we went out for a ride of several hours. We did not have to sell Mr. Chrysler on the virtues of the Airflow ride. It sold itself! Soon Mr. Chrysler showed the same enthusiasm as we did for the Airflow ride, and gave us his approval.

We now expanded our operations. More blackboard work refined the outline shapes. Oliver Clark brought in his assistant, U. L. Thomas, and the rest of his crew to help speed things up. As we came closer to production, we brought in our chassis design crew, Mulhern and Werdehoff, and Carpenter on engines. Herreshoff's preliminary design then was turned over to the production draftsmen.

Birth of the Airflow Car

The prototype Airflow passes over the Rifle River, a popular recreational canoe river near the Mead farm. The deck lid hatch on this prototype gave access to a hidden spare tire which was adopted on coupe models. In sedans, the space created by moving the rear seats forward was used for the storage. Access was through the interior of the car via a hinged seat back, one of the earlier examples of a built-in luggage space. (Breer Collection)

Eventually we built two six-cylinder test sedans and one eight-cylinder model. The body parts were alike except the front end was longer for the eight-cylinder in-line engine. These experimental cars were of even larger inside dimensions than our first test model, and were curved downward over the back even more so than our final production releases. The surprising thing was that the monocoque construction with its non-rigid chassis frame, as released for production, was lighter than our conventional all-rigid frame plus body type. It gave an estimated savings of between 150 and 200 lb.

We shipped these new experimental cars under cover of darkness to our enlarged facilities at Grayling. The contracting advertisers, Lee Anderson and his group, were brought up to date. Lee in turn arranged for a number of celebrities to be taken to Grayling to be photographed with the cars. Among them were the fliers Jimmy Doolittle and Alex DeSeversky, and the famous writer Ernest Hemingway, several of our directors, and many others.

The last step before building a prototype in sheet metal the full scale, wood model version of the Airflow that would be evaluated for its exterior design characteristics. (Breer Collection)

By the time our detail production designers got through, adding a little metal here and a thicker gage metal there, the Airflow became more husky, adding around a hundred pounds of weight. There was not enough time to change all these details when we became aware of this, for we then were headed for production. This additional weight forced us to make a quick decision to add the automatic Keller clutch overdrive.

Production was scheduled for 1934, which meant we had to show our new products at the New York show early in January of that year. This was a rush both for engineering and production schedules. Both Chrysler and DeSoto agreed to introduce the Airflow lines. There was a sedan and coupe in the six-cylinder line for DeSoto, and an in-line eight for the Chrysler. In addition, we made a larger limousine known as the CW, in reality a nine-passenger luxury sedan.

When introduced, the DeSoto Airflow, shown here, was available only as an Airflow model whereas the Chrysler offered two models, the Airflow with eight-cylinder engine and a conventional bodied car with a six-cylinder engine. DeSoto 1934 sales dropped by some 40% over 1933, but much of that loss might have been attributed to a sharp increase in price asked of the Airflow DeSoto over that of the previous year's model. (Breer Collection)

The Chrysler line version of the Airflow. Note the "waterfall" grille that was replaced the following year by a more conventional "prow" type. The Chrysler version of this grille was markedly different from that of the DeSoto. (Breer Collection)

The Imperial Airflow featured two innovations: a curved one piece windshield glass and an electrically operated glass separation between the front and rear seating compartments. A total of 107 CWs were built over the life of the Airflow cars making this a rare model indeed. Only 10 exist today. (Breer Collection)

Birth of the Airflow Car

On the CW limousine we introduced two of our new innovations to the public which quickly were adopted by competition — a curved safety windshield, and an electric, motor driven limousine window lift to raise and lower the glass partition between the front and rear compartments.

Curved glass was new and too expensive to be used except on our luxury CW. Because of the extreme width of the windshield on the Airflow body, we compromised on the standard cars by using two flat safety glass shields projecting forward in "V" fashion with a narrow center divider.

It was 1934 and the New York annual January auto show was now on. We arrived at the New York Palace with a fine array of Airflow show models. Along with this was all the publicity, glamour, pomp, and ceremony that advertising ingenuity could provide. "A new innovation in motoring!" A car "Fashioned by Function" was the byword. "A unit body, structurally stronger and safer, with a lower center of gravity. New weight distribution, along with a lower rate of spring flexure, offering an entirely new ride. New riding comfort! Quieter, and no annoying secondary vibration! The whole car is lower! You now sit lower, forward of the rear axle, and can now lean back and relax, read, or sleep in comfort." Large photographs showing these features were on display and movies showing similar results in action on the road. These cars were an entirely new concept and came as a complete surprise to our competition.

Public acceptance was demonstrated by the fact that more people signed orders for Airflow cars during the show than any other new car ever exhibited at this event. The public who saw them wanted them.

Unfortunately our timing was wrong. Had the New York show been delayed until April when Airflow production was underway, or had we been able to put some 25,000 Airflow cars in the hands of owners during the month of January immediately after the show, we felt certain that Airflow styling would have been accepted. It would have forestalled rumors from developing about the car's reputation.

Meanwhile, production tried to speed things up; new welding fixtures for making body assemblies automatically had to be tuned to operate effectively without delay. The new interior designs — seats with chrome tubular construction — called for added effort. Not a thing went wrong, but the rushed schedule to meet the show delayed the development of quantity production for weeks running into months. It was beyond our power to quell the rumors about the car's quality that developed during the interim, rumors that the delay resulted from all kinds of problems in the Airflow design.

The Airflow received extensive publicity when announced, publicity that often featured the Zeder-Skelton-Breer engineering team. In this publicity shot, Breer points out the problems encountered by the then conventional auto design when exposed to wind tunnel experiments as opposed to the design of the airplane model seated on the desk top. (Breer Collection).

This obviously staged shot of the Airflow in the Michigan woods was meant to convey the feeling of awe that the public would feel when seeing the Airflow for the first time. (Breer Collection)

It is anyone's guess as to what may have happened had some smaller manufacturer introduced a radical new design but could not immediately place enough cars representing it on the highway to make a mass impression. The probabilities are that the odds would be as against its success as they were of the Airflow's.

And the Airflow was rebuffed after its introduction in 1934. Only some 11,000 cars were placed in the hands of owners. As a result of much discussion, we decided to back track on the original concept and change to a forward protruding grill more similar to conventional grills of the time. This was accomplished without a major tear-up for 1935. Even at that, sales fell off by several more thousand. Without further major changes, only minor detail modifications, the Airflow design was continued for two more years. For 1936, the sales ran around 6,000 and tapered below 5,000 in the last year.

The Airflow's negative success might be simply expressed in four words — "Too little, too late," as far as I am concerned, although others expressed later that it was too far ahead of its time. To me, it was not so much that the Airflow was ahead of its time as that Airflow production was way behind time.

Without trying to analyze the pros and cons that resulted in the demise of the Airflow, let me say that although the Airflow as a total vehicle did not succeed, its many innovations were quickly adopted by competition on their own new cars.

In spite of the many new basic innovations introduced by the Airflow, our own corporation executives' interest in the revolutionary vehicle faded, and probably rightfully so.

At this time the Briggs Manufacturing Company threw their styling division into the picture and sold our corporation a bill of goods about going back to the wedge shaped styling of cars prevalent before the Airflow design. Our own engineering division became a divided house. Narrow windshields and wedge-shape front ends again became the order of the day. Of course, some of the Airflow features such as lower spring rates, low rear seat forward of rear axle, weight distribution, etc. were incorporated. Aerodynamics was gone with the wind. Who cared if cars were running backwards as before? Wind resistance was belittled by the competition as having no effect on automobiles; yet today we continue to hear of exploratory work being done by modern laboratories to reduce wind resistance.

Reverting to our former way of building a short body with separate forward hood mounted on a heavy rigid chassis frame allowed greater model flexibility, and helped to reduce selling price. This received the stamp of approval of the buying

Shown here is one of many Airflow publicity shots featuring Walter P. Chrysler in company with Zeder and Breer or with the three plus Skelton. Throughout his life, Breer was convinced that the Airflow did not fail because "it was too far ahead of its time" as critics were oft to state, but "because production was way behind time!" There was an inexorably long delay between the time the Airflow was introduced at the New York Auto Show in January 1934 to much excitement, and the spring of the year when all of the production kinks had been taken care of and production finally went into full swing.
(Breer Collection)

public to the extent that they purchased some 100,000 of our cars during the last year of the Airflow car (1937). Of course, most of that 100,000 were of our conventionally styled models instead of the Airflow. This buying upswing kept on until the government stopped all car production when we entered World War II.

Even though the Airflow had been abandoned, we continued in our aerodynamic investigations. A. G. Herreshoff took on the responsibility of exploring what major gains could be made by reducing the overall wind resistance of our normal size cars by modifying the front and rear end shape of a car. Road tests were run on a straight section of highway parallel to the Michigan Central Railroad near

Grayling, Michigan. The test car was of sectional construction. In brief, a major gain could be made by adding a long tapering tail section to the car. As Herreshoff and associates expressed it, the car, once up to speed with its engine cut off, simply did not seem to want to stop — it just kept on coasting. As I remember, with the long tail section, we could get car top speed stepped up approximately from 65 to well over 100 mph.

So, in the end, the concept of the Airflow led the way to the many fundamental innovations that set the basic design for the modern car of today.

Elimination of the Outside Running Board

Between 1937 and our entering World War II, December 1941, we looked for outside body styling talent. This search brought in Ray Dietrich, giving us further styling independence from Briggs Manufacturing Company. Dietrich's styling theme might be characterized by the word "roundness." Dietrich had an obsession for ample roundness in every manner and form. The longitudinal cross section emphasized curves of roundness. The fenders in profile from the step-in height emphasized roundness. In reality some of us felt that roundness was carried to extreme.

Nevertheless his styling concepts went into production in 1938 with other features and met with public acceptance that year and in 1939 as well.

Late one afternoon I went up to the fourth floor where styling concepts were tried in full-size clay. Oliver Clark and Fred Zeder were going over the finishing touches on the proposed 1940 model for Dodge Division. Just the three of us were there. Everyone else had gone home.

Personally I had always taken an interest and delved freely and frankly into styling, especially after experiencing so much resistance to our Airflow style. Our return to what I considered backward styling, moving back into the wedge-shaped and narrow front end treatment, narrow windshield, continued to bother me. As a result, I guess I still had a chip on my shoulder.

In the late, quiet atmosphere I said, "What's this for, and where are we going to use it?"

"This is the new 1940 Dodge, just about ready for styling approval," was the reply.

I came back abruptly, "I've been watching this job, as you know. As I see it, it is no different than this year's model. On the road you could not tell the difference between it and our present production cars. What's more, by the time it goes into production, its styling will be obsolete. I certainly can't see us going into production with something already obsolete."

This set off a hot and heavy debate using many words emphatically not found in the dictionary, which was nothing unusual between Fred and me.

I continued, "At present the running boards are gone. Why stick to the curved down fenders, just as if we still have to meet a running board that does not exist? Let's raise the horizon line. If I were directing this job, I would clay the front and rear fenders with their tops sloping upward toward their rear end. Make them exaggeratingly tail high. Rear end, tail high fenders will probably look absurd but then we can begin to shave off clay until the moment the fenders look reasonable."

My argument had an effect: a new line of clay models was started. Then after we had the fenders shaped to the way we wanted them, we rounded off the back of the body and raised it accordingly. The result of this compounded curve rear end turned out to be attractive, and created a new era in styling, eliminating the added hump trunk effect at the rear which was quite common at the time. The spare tire was put inside with ample trunk space left when the rear lid was raised. Likewise, the front end was treated in a compounding manner. This new theme in styling was well accepted by the public and more or less carried on until World War II took over, then all car production stopped.

Eulogy for the Airflow

Many consider that the Airflow design, in spite of its sales setback, had laid the foundations for the modern car of today. Following are various magazine extracts printed in the English magazine <u>Autocar</u>, June 19, 1953, titled "Chicken or Egg?"

> "The prominent automobile engineer, Mr. Maurice Olley, now with the Chevrolet Division of the General Motors Corporation of America, has always displayed an admirable directness of expression, and his latest remarks about post-war European cars show that the years have not dimmed his ability in that direction. None-the-less, his summary of styling comes as a shock: 'In appearance, however, nearly all post-war European cars are related to American styling. British and Germans have accepted it perhaps unwillingly, the French with reservations, and Italians enthusiastically.'

Birth of the Airflow Car

"After the first incredulous scoff, the mind searches back for refutation, but it is somewhat difficult to find it. Slab sides and the all-over front with recessed head lamps are as American as the George Washington Bridge, though far less elegant than that splendid structure. Squarely, pugnaciously as a rock in the seas of memory, stands the Chrysler Airflow, a production car which can still be mistaken for a contemporary model although it appeared twenty years ago. It was indeed the egg which preceded the chicken of the modern style.

"Directly to refute Mr. Olley's implication of European subservience, the tendency is to think of the great Italian coach building names such as Pininfarina. But have they done more than impart a classical elegance to the cocktail bar overblownness of the American shape as it first appeared? It is very difficult to make such an assertion. Only in the sports car field can Europe claim to have developed a shape without reference to Detroit. The traditional type is purely functional as a piece of engineering, its modern counterpart for aerodynamic reasons. In these, therefore, "styling" has been a more polishing process. With regret therefore, and even a little chagrin, it must be conceded that Mr. Olley is right.

"In the early Thirties, a talented group of Chrysler engineers and stylists headed by Carl Breer laid down the pattern for automobiles you and I are driving today.

"On the assumption that the average car buyer of that period wanted something entirely original and forward looking, they produced that pattern, the Chrysler Airflow, in quantity and offered it for sale in a variety of models, engine sizes, and body types. They built the proverbial better mouse trap, advertised it heavily, then waited for the path to be beaten to their doors. It never was. After four years the Airflow was allowed to pass quietly out of existence.

"Today, after twenty years, the Airflow is beginning to collect the tributes it should have had in 1934."

It would not be amiss to quote a few lines from Borgeson and Jaderquist's recent book, "Sports and Classic Cars," in which a picture of a 1937 Chrysler Airflow was shown in the Classic Car Section, Part II. Following are extracts:

"While the Airflow is scarcely a classic in the ordinary sense, it is included here because of the tremendous effect it had, not only on Chrysler, but on the industry as a whole.

"Of all the major manufacturers in the Classic era, Chrysler emphasized its luxury cars the least. That is why the firm produced fewer cars of collector interest than Packard, Pierce-Arrow, and others. But Chrysler's unpredictable qualities — as illustrated by the Custom Imperials of the early Thirties and the C. W. Airflow — make it a rewarding marque for study."

Design Guidance of Nature's Fundamental Laws

The exploratory work which led to a better aerodynamic shape as expressed in the basic Airflow design exploded all reasoning and thinking with respect to what was considered standard operation in motor car design at the time.

Moving the rear seat forward some 20 inches to lower the seat thereby eliminating the bulbous ballooning of the body over the rear seat upset everything. The fascinating thing about this change to lower the wind resistance was that all other changes that followed, each in their turn, fitted favorably with nature's fundamental laws.

The following basic changes were made as a result of Airflow theory:

1. Moving and lowering the rear seat rear passengers took them like "humpty-dumpty" away from their precarious position on top of the wall (axle), and also led to a lowering of the center of gravity and a better distribution of weight.

2. Moving the body forward gave us a longer front spring, allowing us to lower the front end jouncing rhythm to a more natural frequency within the human comfort range.

3. Moving the engine forward further away from the car center of mass helped respond to nature's law of the "center of percussion," meaning an upward road force at the front wheels would cause the car body to pivot around the rear axle or at the rear wheels to pivot around the front axle.

4. Lowering the car center of gravity added up to better stability.

5. Building the body as a monocoque, all-steel structure added much to its rigidity, and made it a safer protective unit.

6. The resulting wider front seat and windshield added much to the forward vision of the driver and front seat passengers and improved seating comfort.

7. Other innovations of miscellaneous importance were:
   - The flush type head lights for their aerodynamic contributions.
   - The over-all, one-piece curved windshield for better vision and appearance.
   - A hood lid, hinged and opened from the front, which is standard on cars today.
   - The relocation of the steering mechanism for structural reasons.
   - The addition of a solid rectangular metal roof panel in place of a combination of fabric and wire set flush in the roof for structural rigidity. (Later G.M. introduced the new Body by Fisher with the one-piece stamping all steel top.)

Following is a copy of a letter expressing appreciation for the safety to the driver and family by a 1934 Airflow from an accident caused when the owner-driver fell asleep:

R. Dwight Ware, Pastor
400 East Main
Thomasville, N.C.

September 24, 1934

Engineering Director
DeSoto Motor Car Company
Detroit, Michigan

My Dear Sir:

Recently, while en route to Chicago with my family — three young sons and my wife — my DeSoto Airflow sedan left the highway and was wrecked near Corbin, Kentucky.

I am wondering if I may give you the story of this wreck and a few appreciative comments about the car.

I fell asleep at the wheel and allowed my car to go down a 30-ft embankment studded with rocks half the size of a man and smaller. My car turned over once lengthwise and perhaps a half dozen times sideways and finally came to a stop, top down, with the family all scrambled inside the roof. Except for painful injuries to my wife's arm and a few other minor cuts, all of us were practically unhurt.

Until this year, for eleven years I have driven Lincoln cars and this year also have an Oldsmobile, but I do not hesitate to say to you that I feel that in either of my other cars, and that in any other car designed differently from the Airflow, the kind of wreck we had would have resulted in a much sadder tale. You will see from the enclosed photographs that the structure of the car did not collapse, and that though it inevitably received serious damage, fundamentally it emerged with me feeling that it is the safest car ever built. I had safety glass in every window which, with the girder-scheme in its design, gives insurance against injury not otherwise available.

I do have one or two comments to make that may be of value to you. One: The metal molding overhead in the body of your car came loose and its sharp edges caused painful cuts and makes it a potentially very dangerous hazard (yet I strongly like the kind of roofing or ceiling in the car provided this metal hazard can be removed.) Further: (not observed from the wreck) have a rubber cushion arm rest alongside the doors. Prevent leakage in the cowl ventilator. Improve steering. Make a fastener for the luggage compartment behind rear seat (in a wreck such as ours had not the luggage been packed tight, it would have flung out and all over the children who rode in the rear seat). Increase capacity of glove compartment. Improve window ventilator mechanism. Make keys fit garage padlock as well as all car locks. Lessen turning radius. Eliminate tendency of horn button to stick.

You will note that I have not commented on free wheel. Frankly, I do not know much about it and did not incline to use it much. Furthermore, I did no winter driving and cannot say anything about such. Again, I did not comment on the front appearance of the car. I do not object to this, although I am sure a part of the public do.

I feel so much pleased with the Airflow that as soon as I get an adjustment on my wreck and as soon as the 1935 cars appear, I want another. The thing that entered most in my initial desire for the car was its roomy inside.

I hope you will accept this letter as one that comes from a customer entirely friendly to your product and company and as one grateful for the safety of the car and happy in its use.

Sincerely yours,

R. Dwight Ware

P.S. I am enclosing photographs. They do not show the extent of the damage. A repairman from Louisville came over and examined the wreck and estimated the repair cost to be $500. The position as shown is not that of the actual wreck, because the wrecker men moved it from the deeper part of the embankment to the shallower in order to facilitate its removal. Further, the spare was removed and put on the car, etc. Please return these photographs. RDW.

The steel girders prevent either the radiator or engine from pushing back into the front seat compartment. I have likened this protection to that given the human eye — surrounded by bone "girders."

The Law of Center of Percussion

Another equally important change toward human comfort was the redistribution of weight, more specifically, moving the weight factors further away from the center of mass which helped in two ways: Technically, it noticeably increased the moment of inertia which slowed down fore and aft oscillation; at the same time, it brought the points of center of percussion closer to the front and rear axle centers.

Anyone riding in cars built before the Airflow would remember the arguments about who was going to sit in the front seat and who was going to ride in the back simply because the rear seat ride was terrible. One lived in fear of being pitched into the roof whenever his car passed over a road depression or a bump. In fact, when driving our test cars through the Pennsylvania mountains, we used hold-down straps buckled over our laps to keep us from bouncing so much. Today, hold-down straps for safety in case of an accident are accepted as standard equipment.

I later found, to my surprise, that the law of center of percussion was published in Penberton's View of Newton's Philosophy printed by Palmer in 1728 during Sir Isaac Newton's time. As far as automobiles and buses are concerned, this law was never referred to until we discovered its importance in its application to the Airflow car. A simple explanation is the use of a carpenter hammer. It has no sting if held at the center of oscillation when striking a nail. Similarly, striking a baseball with a bat has no sting if the ball is hit at the bat's center of percussion, while being held at its center of oscillation.

The most interesting part about the fundamental relationship between the centers of percussion and the center of oscillation is that they are interchangeable. Take the DeSoto Airflow four-door for example. Its axle locations with respect to weight distribution were exactly at their respective centers of percussion. The body would rise just as if it were solidly hinged over the rear axle without any upward or downward force at the rear axle. And when the rear wheels went over the same bump a second later, the rear axle upward force would cause the body to rise as if it were accurately pivoted over the front axle. By contrast, in a standard car of the time, when the front wheels went over a bump the body would rise, but instead of pivoting at the rear axle, the upward action would create a downward force on the springs over the rear axle, compressing them, which added force to the upward rebound of the rear wheels as they went over the same bump.

Such changes adopted by the Airflow were quickly and quietly adopted by competition and are the accepted standards of today.

Human Comfort Range

Let me take a moment to expound on some basic thinking that might add import to my discussion on the human comfort range, and the subject of the "Law of Center of Percussion."

When we look at a human being we see one of nature's most wonderful mechanisms, consisting of levers of bone which are directed by the senses and operated by sheer muscular strength. Its legs are used mainly to move from place to place. Its hands are multiple levers in the form of fingers, useful in doing mechanical work in an all-around immediate environment.

We can experience the thrill primitive man experienced when he discovered that he could pry things loose by means of a club and do many things that he could not accomplish without the lever. Think also of the thrill of discovering that if he built the lever in a continuous form about a pivot he had a wheel. What a revelation! Today we mark this accomplishment as one the important points in the advancement of civilization.

And yet at this very moment, engineers wonder in amazement about the present endeavor to adopt man-made laws to limit and restrict automation and labor-saving devices. The cave man would have used better judgment.

Man advanced from the old stone age to the new stone age some 15,000 years ago, and then to the bronze age some 5,000 years back, then to the iron age when civilization began to flourish, and now to the steel age, looking forth to the atomic age.

Taking full advantage of our worldly endowment of elements and using God's fundamentals to our best knowledge, we have, in the rapid progress of civilization, advanced to the motor car of today. When we think in terms of what caused us to develop the motor car, we return to our human desire to move about in a more efficient manner. On water, by dugout canoe, and more importantly on land, by foot and animal. Horses were not dependable and did not fully satisfy our wants, so man made a machine to substitute for the horse, and another step in the advancement of civilization was taken.

Briefly it was first necessary to make it run, second to make it endure, third, to make it quiet, and last, and all important, to make it a comfortable and efficient companion. Before we reached this latter stage, we suddenly awakened to the fact that we had been selling automobiles to human beings and, in effect, saying, "Here is our car. Drive it and you will like it! If it jolts you around, you will soon adapt yourself to its inequalities."

Soon we awakened to the fact that our thinking was all wrong — the human being was the relatively fixed entity and the automobile must be adjusted carefully to suit man's comfort.

We reverted our thinking to the evolution of the cave man and life. We thought in terms of this early man walking from place to place for many generations and we discovered that during this period nature built up a most efficient human machine in proper balance in every respect. This beautiful mechanism is a structure of mechanics that we all admire and respect. Thermodynamically, mechanically, dynamically, and in every way, it is perfect, and we forever continue to marvel about it.

So with due consideration to the human being, we study nature's own action. We learn that over a period of evolutionary time man has established a comfortable stride without fatigue, specifically around 2-1/2 to 3 mph. In terms of engineering, we think of it as 80 to 100 step cycles per minute. Let us call this human comfort range. We notice further that the heart, the kidneys, the liver, etc. are all suspended in the body cavity by means of muscular cords. Nature has established a natural rhythm at which the whole human mechanism seems to relax in comfort. One's eyes tune in with the same relaxation.

The results as applied to the Airflow design were so astounding that one could not help but be enthusiastic about the ride relaxation derived from the change from 120 to 90 cycles.

Animation Versus Law of Center of Percussion

Another analogy regarding the material laws of the center of percussion would be a race horse running with its neck straight out and its jockey in a position to make maximum efficiency of its limbs. The fleetest horse is the one with better balance between mechanical mass and force. If forces are not acting with proper respect to the centers of percussion, energy is wasted, creating inefficiency. In racing, trainers discovered that a jockey could speed up his horse and win races by riding forward to help rather than interfere with the horses center of percussion action.

To verify this natural law with animals I asked Bob Janeway, one of my associates, to check the mechanics of a dog. At the time there was a dog show in Detroit, at which Bob selected one of the fastest dogs there, a Russian wolfhound, a blue ribbon winner at the show. Upon studying the data, he verified that this most efficient dog's legs joined the body exactly at its centers of percussion.

Bus Design Recognizes the Center of Percussion Law

Passenger buses were all wrong in their design. We at Chrysler set about to design a bus to provide the maximum in riding comfort under all loaded conditions, whether fully loaded, partly loaded, or empty. It was a unique accomplishment at the time and set the pattern used by the majority of bus makers today.

Essentially we gave it a torsionally rigid center backbone throughout its length and applied two separate rear drive axles, one ahead of the other. We placed two engines at the extreme rear, one on each side of the center beam, with their transmission connected with propeller shafts to their individual drive axles.

The driver sat forward, operating the transmissions and clutches by remote control. A special third engine was designed with a vertical crankshaft to rotate an air propeller which forced the air downward and through the vertical radiators facing to the rear.

The chassis was built and thoroughly tested. It demonstrated the importance of weight distribution. Before the bus body could be built, the corporation decided to discontinue the passenger bus business.

Today rear engine buses are the order of the day, and cannot I help but note the sudden change in passenger bus design following our publicizing the law of the center of percussion because of the great gains in riding comfort provided by proper weight distribution.

With the exception of the double deck buses New York and London, buses previously were more or less adaptations of a truck chassis. Bodies were put on in a sort of "humpty-dumpty" fashion which also typified the ride. Engines were placed forward mostly, just behind the front axle, under an accessible hood. Then came the driver's cab and body, all centered over the rear axle. These are obsolete compared to today's modern monocoque structure with drivers up front and the engines in the rear, and more importantly with the front and rear axles closer to one another under the body structure, thus providing a weight distribution that carefully respects the law of center of percussion.

Racing Car Design and the Center of Percussion Law

From the very start of the Grand Prix by the Automobile Club of France in 1906, the development of the racing car seemed to be mainly by trial and error. The racing car chassis more or less followed that of established passenger car design.

Not long after the introduction of the Airflow and its much publicized original features, racing cars picked up the basic Airflow design trend. One of the outstanding examples was the stepped up Mercedes-Benz design evolving from 1934 onward up to and including the present time.

The 1939 W-163 models, for example, had less acceleration than the 1937 W-125 models but were superior in stability and braking, and may have been the fastest cars in the world on a road circuit. During 1937-9, Mercedes engineers made every attempt to place the weight of the car equally on the front and rear axles, and to mass the weight at the front and back so as to give the largest possible moment of inertia.

After World War II, an even more advanced Mercedes-Benz model known as the W-196 was brought out to comply with the new 2.5 liter formula. Its engine and chassis design, and the bold introduction of aerodynamic bodies with fully enclosed wheels, caused a sensation. It was said to have opened up a new era in the design of the Grand Prix car. It had a frame of a lightweight tubular truss construction. In order to achieve the maximum polar moment, the masses were designed to be at the extreme ends. At the forward end, two large diameter tubes were placed just behind the plane of the front hubs. The steering box, steering connections, and front wheel inboard brakes were ahead of and fastened to the cross tubes. The engine mounting also attached as far forward as possible to these same tubes. The inboard brakes applied their retarding forces through means of universal joints and shafts that angled rearward to their respective front wheels.

I have covered this racing car at some length because of its reputation as one of the fastest racing cars the world had seen, yet its design conformed to the seven basic objectives outlined below:

1. Minimum frontal area with body shaped for lowest wind resistance (wheels enclosed).

2. Maximum forward and aft weight spread (approaching centers of percussion requirements).

3. Maximum horsepower per pound of car weight.

4. Lowest possible vehicle center of gravity.

5. As near as possible equal front and rear wheel loading.

6. Minimum of unsprung weight combined with low spring suspension rate, hydraulic dampened.

7. Maximum braking well cooled for elimination of fade out.

Of interest is the fact that these points also summarize our early findings with the Airflow car. As Al Fisher, one of the brothers and engineer in charge in the buildup of Fisher Body fame, said to me one day, "Your Airflow car was certainly away out ahead in automobile design. You were just too far in the lead to be accepted at the time and it took some time for the industry to catch up with your thinking."

It was interesting to note that the final Mercedes racing car body had covered wheels. Our early wind tunnel findings told us that wind resistance was lower when the fenders were brought down over the front wheels. At the time of our findings fenders were still straight out forward, in some cases turned partly downward. Our findings created interest in future designs based on closed-in wheels.

After World War II our body design division designed and built sample cars for production with Airflow characteristics. We made full-size wood models styled with slightly curved fenders covering the wheels as cars of today. The hood valleys started from the front between the radiator and front fenders, and faded out at the windshield instead of continuing on to the rear around the windshield "A" posts. This proposed design utilized a maximum width of windshield like the Airflow. Unfortunately, cost analysis put the design into a prohibitive cost bracket. As a result, the design had to be discarded, and top management forced a complete redesign which made our line of cars comparably smaller than competitions'. To cut cost of materials, the bumpers extended only to the fender sides. The fenders at the rear had indentations so that the bumper ends could hug in close.

Glass manufacturers at the time had set prices for plate glass sizes. Windshield widths would depend upon the standard glass sizes available. As I remember, windshield widths were set in overall increments of three inches. This is one reason why rear windows were made small. It also meant that the large Airflow type windshield could become cost prohibitive.

Unfortunately, Chrysler Corporation's popularity suffered from its persistent decision to continue using flat windshields, whether two-piece V-type or flat one-piece. This occurred in spite of the fact that we were the originators and first users of curved windshields. Our later change to curved glass windshields after competition had been using them for several years made it tough on the used car market, resulting in a forced sales slump because of the customers' adverse reaction to used vehicles with flat windshields.

There were other reasons why top management and production heads opposed Airflow styling and engineering. One was the fear of being able to manufacturer the slightly curved stampings and still produce acceptable bodies. Of course, bodies today are lot less curved.

Part VI: Railroad Ride Research Along Airflow Principles

Having a home in California I found myself making many trips back and forth on the finest train equipment. I could not help but notice how terrible the ride always was and what a beating one took on our finest transcontinental trains regardless of how smooth and straight the rails were. One could not read for very long because of the jolting and uncontrolled bouncing amplified by the slightest upward movement from the rail. From an automotive engineer's viewpoint, these forces, added to the locomotive drawbar lashing forces, made the ride ridiculous.

When flying out to California some 20 years ago on an emergency call in a DC3, the contrast between the vibration free airplane flight as compared to rail travel was so great that I wrote a letter to Walter Chrysler about it, but I told him to save the letter and compare it with the one I would write about the train after taking it back. I knew that Walter Chrysler, being a railroad man at heart, would be interested. Coming back east on the Santa Fe Super, I had the expected difficulty in writing a letter legible enough to read.

Upon arrival back at the office I gave Bob Janeway 50 dollars of my own money to buy a Lionel electric train set. The toy trains were replicas of the standard Pullman train at the time and I wanted to find out how much faster and safer the trains would operate if their weight distribution was near to requirements of the law of centers of percussion. With corrected weight distribution, we found they would run the circuit track much faster without leaving the rails. The experiments were so encouraging that we contacted Walter Chrysler and told him what we found. Unlike how our automobile engineering was ironing out highway roughness, the railroads to the contrary amplified every force and movement that came up from the rails. Also, engine drawbar forces brought in lateral whipping motions that were very aggravating to the passengers.

Walter Chrysler, being on the New York Central Board, was very interested. He immediately called Central's vice president in charge of the Rolling Stock Division, who was not in, and in his absence, contacted the engineer in charge, Mr. Paul Kieffer. Paul later spent an afternoon with me and as a result we built up a very friendly engineering association in working on the train ride problem.

Our laboratory work was set up on miniature scale. With various setups we could demonstrate the many forces and their play in affecting the ride qualities of the standard train equipment in use at that time.

Shortly after starting our miniature train laboratory, Lionel brought out a toy articulated streamlined train having four-wheel articulating trucks. It was a reproduction of the advanced railroad passenger car service of that time. The entire train was articulated except at the rear end of the locomotive, which to us was getting away as far as possible from nature's basic law of the center of percussion. Speed tests on this miniature train demonstrated that the ride on this setup was noticeably worse than the standard non-articulated trains. In our laboratory we showed marked improvement in ride by using two separated pivoting points on the articulated trucks. This radically redistributed the weight in the car bodies and demonstrated important gains in ride comfort as well as in a much higher speed at which the train could be safely operated. This improvement adhered to Sir Isaac Newton's "Law of Center of Percussion" as expressed in Pemberton's book of Sir Isaac Newton printed in 1728.

Two adjacent cars riding on an articulated truck between them can be made to comply with the law of center of percussion by properly locating separate trucks inward from the outer car body ends in accordance with the body weight distribution.

Our railroad laboratory was most interesting with its many miscellaneous miniature setups. We might enumerate some of them just as a matter of interest. For example, we could show how standard train setups developed "crack the whip" forces, and how these could be overcome with proper weight distribution. A similar setup illustrated the more severe problems of the articulated train. Another demonstration set forth the gains to be made from self-banking cars when going around turns, or how rail forces and truck shock absorber forces would come up into the car body via vibration, sound, and amplified motion.

Our train research and laboratory was headed by R. N. Janeway, with Bill Vandersly and Perce Best to cover engineering and design. Associated with this, we had a small machine shop for instrument and model making.

Many train executives were astounded at our approach to their ride problems. Among them were top men such as Bill Jeffers of Union Pacific and Mr. Clements, President of Pennsylvania Railroad. In fact, Mr. Clements once said, "I have seen more engineering in this laboratory than I have in my entire life in the railroad business."

We set up a test run program on regularly scheduled trains between New York City and Utica, New York — a distance of over 200 miles. The recording instruments used for comparing the progress the many train runs made one to the other, were the "Minor" type recording accelerometer and the "Grey," a type that counted the number of various fixed degree accelerations encountered during each run.

The Grey instrument was adapted with modifications to record both vertical and lateral forces. These forces were recorded inversely to measure passenger comfort. The greater the force the less comfort. The instrument for measuring lateral forces was kept level automatically in order to get consistent results. The instrument measuring vertical was placed directly on the floor. In both cases the instruments were located adjacent to the truck centers of the respective cars.

The recordings seemed to be dependable until sudden snowy weather showed surprising results. The instruments simply recorded that under snow and rain the ride was improved because of water lubrication. Snow worked into every crevice, reducing friction wherever dry surfaces formerly rubbed. When old time railroad men were asked about this phenomenon, they said it was due to "slippery track."

When we examined the train trucks, we found that the heavy leaf springs were dry and rusted, and showed no spring movement at the leaf ends, indicating little if any spring action. The car end passenger diaphragms and buffers also were dry and under high pressure. The bolster rub plates were dry metal-to-metal. When the brakes were applied, the added friction made the entire car shudder. When the spring leaves and all rubbing surfaces were lubricated, there was a marked improvement in the ride. However, this lubrication would not last. Wind with road dust soon would bring back friction. Metal leaf spring covers were applied to maintain lubrication uniformly over a long period of operation. Ball bearing thrust plates were installed between the bolster and truck frame to eliminate friction too, but although they helped, they were not satisfactory. This led to the development of metal thrust links with flexing rubber ends which soon were adopted universally as standard equipment on all train passenger trucks. Then by removing the thrust wear plates, sufficient clearance was obtained to allow both vertical and lateral freedom of action when the brakes were applied. All of these features were carefully covered by our patent division.

Then we went back to our original objective of evaluating the effect of load distribution to meet the law of center of percussion. The cars were heavily loaded with sand bags, and the ride and whip conditions noticeably improved. Roller drawbars were made which reacted as if the drawbar forces of adjacent cars were being pulled from the center of their respective trucks. Later drawbars were made to actually apply from the center-to-center plate of these same respective trucks. The result of these two means were similar. Various laboratory models were built to show how those improvements could be incorporated in a practical design of the full-sized truck.

Timken Roller Bearings were tried in place of plain wheel bearings. The free lateral floating movement allotted with the standard plain bearings showed better ride results than the non-float Timken type.

About this time the first new Cleveland-run "Mercury" cars' interiors had been completed. A last call was made for what little could be accomplished to give them a better ride. The quick changes we recommended for a better ride were:

- Equip the trucks with softer springs (lower rate leaf springs).

- Add spring covers if they could be had; if not, keep the springs lubricated.

- Use high finish bolster pressure wear plates and keep them lubricated.

- Lower the diaphragm and buffer pressure between adjoining cars and lubricate their rubbing surfaces.

- Add thrust links on the trucks (but like the metal spring covers they could not be had in time for the first Mercury. These were later added to the second Mercury running to Chicago.).

The first Mercury official run between Cleveland and Detroit resulted in a marvelous showing. The Mercury was considered the best and most comfortable riding train by far, above any train New York Central had, in spite of its stepped up speed schedule. The Waugh rubber cushion drawbars dampened out the locomotive drawbar or endwise pull variations of the train. A second Mercury-to-Chicago train also was very successful.

While we were doing our train ride testing, the trend in train design took a turn toward lighter weight streamlining. At the same time there was a surge in shortening car lengths and mounting the adjacent ends on a common four-wheel truck, namely, providing articulation.

Several all-steel partly articulated trains had been built when a lighter weight, aluminum train with all cars articulated was announced by the Union Pacific combined with Pullman. The train was named the "City of San Francisco."

Front page publicity covered its first scheduled cross-continent run from San Francisco to New York City carrying top railway and public relations officials.

The very next morning after the run was made, Messrs. Muzzy of Pullman, and Burnett and a Jabbleman of Union Pacific made an early call on me. Muzzy wanted us to discuss our laboratory findings about centers of percussion.

We showed them the effects of weight distribution on the previous trains with cars that had two four-wheel trucks. Then we showed them the effects of articulation. I further illustrated the effects of center of percussion forces when applied in their simplest form. A simple experiment effectively simulates the reaction forces that come through the passenger body from the supporting trucks on the rails below. I took two sticks of mahogany about 2 in high, 1 in thick, and about 14 in long. On one the centers of percussion were marked at equal distance from each end. On the other, in which I simulated articulation on the truck, the center location was near one end, and the natural center of oscillation or center of percussion (as they are interchangeable) near the other.

The three guests then each took a turn resting the stick on his finger at the extreme end. Then I would strike upward under the other extreme end. It would take only one blow to cause the downward force to register a severe sting. But when I struck the stick at the center of percussion not the slightest sting would be felt.

The three sat there for a moment, tired but glum. All of a sudden Muzzy spoke up to his two associates, "Now do you see what I mean?"

Then he related to me what happened on their much publicized fast, cross country run in the articulated aluminum train loaded with top railroad officials. Muzzy tried to sleep in his upper berth but spent most of the time hanging on to the emergency cord which he was ready to pull more than once because he thought the train was ready to jump the rails.

Mr. Muzzy, as head of Pullman Manufacturing Company's engineering, was always enthusiastic about our work. While we were designing a very light truck with welded steel stamping side members for him, Mr. Muzzy passed away. Other Pullman representatives who took Muzzy's place lacked a sense of appreciation for what we desired to accomplish.

Nevertheless, we progressed from our simple, lightweight, stamped steel side member railroad passenger trucks to a second and third design. A set of the latter were run in regular passenger service by New York Central for over 100,000 miles. This set of railway car trucks provided satisfactory service and a far better ride than any other standard truck.

Then the Pennsylvania Railroad Company also became interested, and we developed and tested a pair of new modified trucks for them on a special test train. A number of runs were made up at over 100 mph. We had a favorable ride with one exception: at certain high speed we ran into what is termed as "hunting," a result of cone-faced car wheels fixed on a common axle. The right side wheel would tend to run ahead of the left side wheel by moving up on the larger cone diameter, then, in turn, the left wheel would run ahead. This caused the railway car to swing from side to side. We had made only one test run when the Pennsylvania had reorganization and financial problems and did not follow through any further.

Still, Janeway and our laboratory developed a very interesting test fixture to explore the various factors of the modified design to determine the cause and effect that excited this condition.

Since financial conditions with both the New York Central and Pennsylvania Railroads were not favorable to continue, we took inventory, and decided to develop a freight car truck that would give those cars a ride more favorable than existing passenger railway cars. There were several reasons why we did this. One was a discovery that there was a shortage of freight cars when United States entered World War II. Freight cars at the time had a set upper speed limit above which it was thought that they would jump the tracks. However, shipping capacity could be enlarged by a ten-mile increase in speed. To do so, the already almost rigid coil springs were further stiffened so that the natural frequency of jounce would be raised, thereby increasing freight train speed around ten miles. As a result, freight handling capacity went up some 15 or 20%. Of course, added lading damage in emergency was of little concern. Unfortunately, wreckage rate also went up. The stiffer spring and greater speed combined with the crude few friction dampeners used at the time materially increased fatigue breakdowns, especially causing a breakage of axle shafts. To examine the problems, Timken Bearing Company under the able guidance of Mr. T. V. Buckwalter, a former railroad engineering employee, assisted by the cooperative financial backing of the Associated Railroads, built some full-size wheel and axle shaft assembly fatigue testing machines. Test results indicated that the axle shafts always broke in the same location, at the inside edge of the wheel hub. The high inward pressure of the wheel hub, because it was press-fit, caused a "notch effect," that is the sudden change of strain or metal stress was like putting a notch in a broom handle at a concentric point where it would easily break. We applied the same cure to it as we did to passenger cars and freight car axle breakage became a thing of the past.

Mr. Buckwalter, who pioneered Timken Roller Bearings in the railroad business, later told us how he followed our ideas and cured the Pennsylvania Railroad's troubles with their electric locomotives which would jump the track above a given speed. Pennsylvania Railroad had become enamored with electric locomotives. However, they had to restrict their top speeds to avoid wrecks and delays which materially affected their schedules. Buck suggested they change the locomotive weight distribution according to our Airflow findings with respect to the law of centers of percussion. The design engineers had been locating too much weight on the central driving wheels, and not enough on the "leader and trailer" trucks. The simple change spelled success for the heavy investment which might have otherwise been considered a failure.

The sum result of the New York and Pennsylvania railroad financial problems was that Chrysler Corporation was assigned all patent rights to the development work accomplished to date, and we concentrated on the development of our railway freight car truck. The last I looked, we had equipped some 1,500 or 2,000 cars with these Chrysler railway trucks which were manufactured by Symington-Gould on a royalty basis.

We cannot help but reflect back that our experience in motor car research in designing our Airflow cars had put us in the unique position of providing useful guidance to the railroad industry. Our railroad research experience points very strongly to our conviction that the railroads are missing a bet by not starting with a new slate in their engineering approach.

The program is simply this — start from the bottom with just the two smooth ribbon like rails with the curves as found on all present well-kept railroads. Then build up from there, whether it be the Hildago type of train or the regular, multi-two truck per car train. The following points might be considered:

First — Design a chassis arrangement in which all drawbar forces are applied to the wheels through the trucks, such that all lateral force components caused by drawbar pull or brake drag are taken directly by the rails, thereby allowing each car body to function independently above the chassis.

Second — Spring mount the car bodies on the chassis, locating the spring positions as near as possible to the car body "centers of percussion."

Third — Use coil spring mountings with a spring rate that provides a frequency no greater than 85 to 90 cycles, and for damper control, use friction or hydraulic shock absorbers (Chrysler friction type are now used by Union Pacific for passenger car control in preference to hydraulic).

Fourth — Use a torsion bar spring suspension in place of vertical coil springs to eliminate the annoying secondary high frequency vibration of the spring coils.

Fifth — Use long radius connecting links to control fore and aft movement, also similar links for control of lateral forces in order to eliminate interfering static rubbing friction forces when accelerating or applying brakes.

Sixth — For maintaining easy access for passengers moving from one car to the other, use low pressure, flexible, inflated, tubular, tire-like bellows to seal the passageways between cars so that passengers can move from one car to the other without adjacent cars being affected by the severe lateral or vertical friction forces present in passenger car equipment. This type of air flow could automatically be held inflated from the air brake system and be thoroughly sealed against dust and dirt leakage.

Seventh — Provide centrifugal neutralization for the passengers when going around railroad curves. All railway tracks are banked at the curves; the fixed angle banking of each curve calls for a correct mph car speed to be held when travelling the curve to provide perfect passenger comfort. To take care of variations in speed above or below the correct speed for the curve, it is important to design the so-called chassis centrifugal neutralizing mechanism between the car body. Such a setup we designed and demonstrated in our railroad laboratory. This can be accomplished by arranging the body lateral motion suspension system so that the center of mass relative to gravitation forces of the car body will automatically bank correctly to provide maximum passenger relaxed comfort, duplicating that of flying in a modern passenger plane. Another way to accomplish the same car body tilting result would be by air or hydraulic forces controlled by means of a "leader valve" as directed by gravity versus centrifugal force pendulum action.

There are three different under-chassis types to be considered on which to mount the railway passenger car: (1) the Hildago arrangement as developed in Spain — a chassis with a single pair of wheels at the rear, perhaps unsprung, and the forward end mounted on the next car chassis ahead, pivoted thereto as close as possible over center of the axle of the respective chassis wheels; (2) by applying the articulated train design in which the chassis would be mounted on four-wheel trucks located or centered midway under and between the ends of the adjacent car bodies; and (3) mount each car chassis on two trucks located as near as possible to the centers of percussion of the car body above. This last would have the advantage of being able to travel over the same rails at higher speed with less side rail or vertical aggravation because all forces are applied adjacent to the center of percussion location with respect to each car mass action.

Part VII:
The Chrysler Engineering Team and the War Effort

When war was declared, the armed forces immediately contacted many manufacturing companies to expedite the manufacture and improvement of available armament and fighting equipment.

The armed forces solicited our help in some 1,150 engineering development projects by the army ordnance alone supported by the expenditure of some $18,000,000.

Skelton, in close association with Lt. Col. Joseph M. Colby heading the development section of the Detroit branch of the ordnance division, did an outstanding job in designing and building some 38 new type of pilot tanks including those with long range 8-in bore guns.

The original A-I tanks were immediately scrapped and the 28 ton M3, a complete redesign, went into production. This was replaced by the 32-ton Sherman M4 model which in turn was replaced by the 43 ton Pershing tank. Had the war continued, a tank of 65 ton, over twice the weight of the original, would have been in production.

With reference to all the war material produced by the various Chrysler Plants, K. T. Keller employed Wesley B. Stout, associated with the Saturday Evening Post, to collect the material which was issued in the six books listed as follows:

1945 — "A War Job Thought Impossible"

1946 — "Tanks are Mighty Fine Things"

1946 — "The Great Detective"

1947 — "Great Engines and Great Planes"

1947 — "Secret"

1949 — "Mobilized"

These books are very well written and descriptive of the equipment designed and manufactured, well illustrated both in color and black and white. They cover the visiting armed forces personnel as well as the celebrities, and pay tribute to engineering as well as production.

The armed forces technical men soon recognized the fine qualifications of our engineering setup and the willingness on our part to split into smaller self-contained special groups best suited to expedite each special army project.

As a result, some 25,000 tanks were built, as were 18,000 Wright 3,350 cu in air cooled radial engines, 60,000 Bofors guns plus 60,000 extra barrels for replacement, 5,000 B29 fuselage assemblies, some 300,000 rockets, 435,000 army trucks of many designs, and some 3 billion rounds of small arm .45 caliber ammunition.

Naturally when World War II came on, it was fortunate for both the Chrysler Corporation and the armed forces that we had a wonderful engineering and laboratory team setup which could quickly solve many of the problems that were associated with the development and manufacturing of various war products needed by the armed forces. The decentralization of our various engineers and research men certainly expedited a desired saving of time.

As one example, production accepted a major contract to build a plant in Chicago to produce a large number of 3,350 CID 18-cylinder, air-cooled, carbureted, twin-row radial engines to be used in B-29 bombers and other fighting planes. The engine had been developed by the Wright Aeronautical Engine Company in Buffalo, New York. As soon as the contract was signed for Chrysler to build a plant and manufacture the engines in Chicago, we set up a separate, self-sustained engineering division in our Detroit Dodge car plant to explore, check drawings and parts lists, and help the design layout for the Dodge Chicago plant testing facilities.

Our engineering group, along with production, soon discovered that the Wright Aeronautical engine design was not controlled by drawings. When several engines were torn down and their parts mixed, they could not be reassembled because the parts would not go together. This indicated that the Wright engines were built by craftsmen, and either the drawings were not exacting or the mechanics disregarded their plus and minus dimension limits.

Chrysler was justifiably proud of its contributions to the World War II effort, so much so that K. T. Keller later authorized six books to be written of this period in its history, one each for the years 1945 through 1949. Scenes such as this meeting between Chrysler management and the Army in review of the T93 eight-inch gun carriage were not uncommon. (Breer Collection)

Our engineers asked us to contact the Wright group to clarify the situation. In order to save time and the argument of having us as a third party go-between, we advised our men to setup an independent engineering team and work directly with the Wright Engineering organization. We would supply them with all the technical and laboratory help they needed.

This engineering team in cooperation with our laboratory organization made many physical and metallurgical refinements in the engine from its very start. The original engine was hazardous because several of the rear radial row of cylinders would overheat due to cooling air flow around the rear cylinders being blocked by the complex carburetor manifold.

In the same large dynamometer laboratories at Wright Field in Dayton, where endurance tests were being made on the released 3,350 carbureted engine, fuel injection was under development for the same basic 3,350 engine. The simultaneous testing brought out the fact that the carburetor production test engine had to be torn down and overhauled many times more often than the same basic engine upon which fuel injection was being perfected. This forced a quick decision between Wright Field and our Dodge Chicago engineers to perfect the full injection system as quickly as possible. Our group in Chicago was credited with this great accomplishment, making the fuel injected 3,350 cu in engine the most dependable, outstanding, reliable aviation workhorse of that day.

In fact, after the war this fuel injection engine became very popular in commercial transport aviation.

Development of Tanks and Special Types of Truck and Vehicle Equipment

The war effort also saw Chrysler set up a much larger design and engineering section for the Ordnance division to design and develop tanks and portable quick-acting field artillery.

Chet Utz handled the design section in close association with Engineering's Fred Slack, cooperating with Skelton and Colonel Colby. Elmer Dodt was sent to Aberdeen to facilitate design decisions with Ordnance engineers early in the tank's production development.

At the Chrysler Detroit Tank Arsenal, a production type tank test track with its dips and mounds was immediately set up just outside the Arsenal manufacturing plant.

As all auto production had been stopped, the Packard Proving Ground test track and hill surroundings in Utica in the outskirts of Detroit were taken over, and a grueling test engineering organization was set up at its location. There Tobe Couture, Frenchie Raes, and others did a marvelous job in field testing to increase the endurance life of our production tanks, as well as add many new engineering improvements in tank truck and track laying suspension.

In addition to this work was the testing and waterproofing of many types of army trucks, transport vehicles, and all special four-wheel drive vehicles.

At the start of war the truck and transport vehicle requirements kept that branch of our manufacturing plants very busy. A special design group involving a large division of designers worked closely with the arms division developing special vehicles, both two- and four-wheel drives, in all forms for transport for troops, such as swamp buggies or special Red Cross ambulances.

For example, under extreme secrecy, our engineering designers cooperated with the Signal Corps to simplify the gear-driving mechanism and build into a single 19-1/2 ft semi-trailer the radar electronic equipment that in pre-war times took seven trucks to carry. The new van was produced by Fruehauf. When set up for action, it held a large radar scanning reflector and rotating mechanism raised up through the roof center of the van. A series of self-contained leaning jacks rigidly supported the vehicle and kept it level. As a result of this cooperative effort, Dodge Division delivered some 2,000 complete Radar antenna mounts.

Another fine example was our engineers' and designers' adaptation of the Swedish Bofors rapid fire automatic cannon to many purposes for the army and navy. These guns could fire up to 140 rounds of 40 mm shells per minute. Some 60,000 of them were built plus another 60,000 additional gun barrels as spares.

Twelve corporation plants participated in the making of parts yet the major portion were made in the Highland Park, Plymouth, and Jefferson plants.

This beautiful but rather complicated mechanism, the Bofors, was made mainly by hand craftsmen and was said to take some 450 man-hours of bluing, scraping, and filing to assemble. Through the cooperative effort of our designers, engineers, and laboratories by making an accurate set of drawings, substituting inches in place of millimeters, and by using castings plus many parts of oilite or super oilite, assembly time was reduced from 450 to 10 man-hours.

As a sidelight, our stress analysis laboratory increased the brief life of the shell extractor level. The average life reported from the field was only 1,500 rounds. The natural tendency would be to increase the extractor lever dimensions, but we made a replica of transparent plastic, studied it when stressed under polarized light, and soon determined that the original design had a localized stress that could spread over a wider area by removing metal in an adjacent area. As a result, a new lever design was able to last three times as long as the old.

Let me also credit Chrysler engineers for coming up with solutions to other problems not involving vehicles. For example, our Bob Christman developed a special steel to meet cartridge requirements replacing brass which had come into short supply. We produced 3 billion rounds of 45 caliber and 500 million rounds of 30 caliber cartridges. Also, Dave Wallace from our Jefferson plant designed and produced some 29,000 marine engines and 34,000 industrial engines to be used in mobile fire pumps, air raid sirens, and smoke screen units. To these we might further add 9,000 pontoon units that could be assembled into large, multi-unit barges. The list can be embellished again by adding the manufacture of about 320,000 rockets as well as multiple rocket launchers for use on tanks and trucks. Finally, the Chrysler Air-Temp Division manufactured 15,000 air conditioners and refrigeration units to protect food and medical supplies, and delivered some 62,000 field ranges, over 17,000 furnaces, and 30,000 heaters for troop use.

Multi-auto Tank Engine Designed and Developed by Central Engineering

One of the major challenges that our engineers accepted was to expedite the design and development of a substitute tank engine to replace the radial aircraft engine which was in short supply because of aircraft demand.

Our reaction was to place five of the dependable 1942 six-cylinder engines used in the Chrysler Royal and Windsor models arranged in a radial engine fashion but operating in normal upright positions with all crankshafts running parallel. Each engine delivered its power through herringbone gears to a center power drive shaft. This made a compact power unit of some 750 hp in the same space required for the air cooled radial engine without a major tank change. As soon as we had this arrangement on the dynamometer stand checking for power and performance, Bill Knudsen, then headquartered at Washington, came over to see how we were doing. We took him and his associates to our dynamometer lab. After seeing it perform, Bill took me to one side and asked, "Carl, do you think it will do all right?"

I told him that we had been building thousands of six-cylinder engines over a period of years. Their wide-open power endurance tests had demonstrated a dependability well beyond anything expected in the most severe car life tests, and that they were more dependable and durable than any air cooled aviation engine design we knew about. The only problem they presented was the matter of gearing, and the relative reaction of each engine to the other. Some 50,000 of these automobile engines were put through Chrysler production lines and assembled into multi-engine power units for tank use. The ordnance division shipped most of the tanks so equipped to England and nothing but the finest reports came back about them. The English swore by them, and wanted none other for their use.

Sixteen-cylinder 2,500 HP Aircraft Engine

One of the largest research projects that we took on was to develop a 2,500 hp liquid cooled aircraft engine for the Air Corps at a cost of about $14,000,000.

Our manufacturing division was called in to look over the Rolls Royce 12-cylinder aviation engine in production in England with the intent of putting it into mass production. Our production people were not keen on tooling up for an engine not of our own design. As a result, B. F. Hutchingson arranged for a party to discuss this Rolls engine with General George Kenney, the directing head of Wright Field at Dayton, Ohio, and under whose direction all major development work of planes and engines was handled.

General Kenney told us that he had just returned from France and noted that he saw many planes on the ground inactive. A French general said that when they go up after the Huns, the latter just fly away from us. If we try to follow, they turn and simply riddle us. They were too fast. What he needed was an airplane engine that puts lots of power through a small hole in the air.

I told Kenney that the Rolls was a V-12 cylinder and perhaps a V-16 might be more effective. Kenney replied that they had enough trouble with the V-12 because of the length and torsional weaving of the crankshaft. I replied that the Rolls had a reduction gear at the front end. If we were to put the reduction gear in the center of the engine we could place two V-8s back to back driving the reduc-

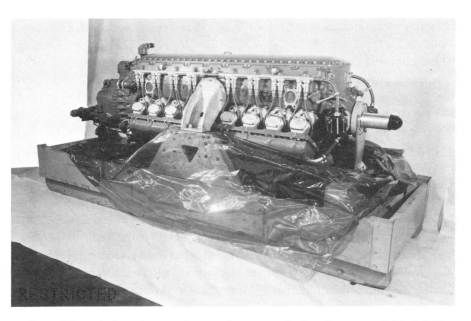

One of Breer's favorite wartime projects was the development of this 2,500 horsepower, 2,250 cubic inch aircraft engine comprised of 16 individual cylinders bolted to a one-piece cast aluminum crankcase. It underwent one very favorable test flight, then was shelved by the Air Corps as the war ended. (Breer Collection)

tion gear for the more efficient lower propeller speed. Kenney agreed and signed a contract to develop a 2,500 hp engine having 2,250 cu in displacement and weighing 2,430 lb.

It was a tedious job. The design was composed of 16 individual single cylinders bolted to a one-piece, cast aluminum crankcase that we hoped would have the fatigue strength of German aviation engines already in production.

One of the first problems we encountered was the fact that the German engines had a cast aluminum crankcase fatigue strength according to our laboratory tests of around 21,000 psi. The best that the Aluminum Company of America could show was about half that.

We called in the Aluminum Company of America engineers, showed them our problem, and told them that if our engine was to go into production, it would call for more aluminum per engine than any other now being used by our air corps.

We pointed out that we wanted an aluminum metal casting that would give us a fatigue strength of 21,000 psi, and asked for their help. They said, "Fine," went back to their headquarters in Pittsburgh, and they never reported back.

The upshot was that our own metallurgical unit went into their foundry, and, in less than a few months, took over its operation. By careful control of temperature, cleanliness, purity of metal, and quantity of excess hot metal flow to establish the proper balance of uniform conditions in the mold, we were able to develop castings with a fatigue strength equivalent to the German fatigue strength accomplishments of 21,000 lb.

Unfortunately, contracts for this engine never materialized. First George Kenney was transferred, then his replacement, and it had lost its champions. Nevertheless, our engine was completed and flown in a single ship. A pilot named Cushman took up a plane with one for the first time, and was so enthusiastic in the way it handled that he came over the field at low altitude and made a barrel roll. This was against all orders in the routine of initial development testing. When he landed the field officer reprimanded him and issued orders for a complete teardown of the engine for inspection before flying again. This was the one and only flight made of which we knew.

The engine delivered 2,500 hp with a displacement of 2,250 cubic inches, more than 1 horsepower per cubic inch. With a weight of 2,430 lb, it provided more than a horsepower per pound. It was shelved as the war ended because the air corps engineers already were concentrating their attention on turbo jets and jet propulsion.

1,000 HP Gas Turbine Development vs. Car Gas Turbine

After the war, a multi-million dollar contract was signed with the Navy air corps to develop a turbine driven propeller power plant of 1,000 hp with an economy equal to or better than that of the reciprocating engine. This involved the development of a lightweight heat exchanger of unusual capacity and efficiency. All specifications were met. The turbine engine-driven propeller was test flown. Changes in the navy organization as well as their outlook on future planes stopped further work on this project. The engineers working on the engine were under the very efficient supervision of my research division assistant George Huebner. He later took on the development of a gas turbine power plant for motor car use with the objective of higher performance and the same or better economy and first cost of a piston engine. Some years ago I had my first car ride with the gas driven power plant.

Part VIII:
Death of Walter Chrysler and a New Regime

Walter Chrysler took sick suddenly in 1938, and passed away in 1940, a year before Pearl Harbor.

We missed his good counsel and his fine cooperative thinking. Our close association with Mr. Chrysler was a very happy one. He was the personification of the corporation named after him.

Chrysler had a lot of respect for our engineering ability, and expressed so in his book, "The Life of An American Workman." During his regime advertising and publicity centered around our engineering innovations and achievements.

With the passing of Mr. Chrysler, Mr. K. T. Keller stepped in and took over the reins. Keller had been with General Motors at Buick in Flint, Michigan. He then became vice president of Chevrolet, and later the General Manager of the Canadian Division of General Motors. Walter Chrysler hired him to become manufacturing manager of Chrysler Corporation in 1926.

Keller had with him his very able manufacturing associate, Herman Weckler, a University of Pittsburgh graduate, who had originally joined the American Locomotive Company where his father had an important responsibility. Weckler directed purchasing engineering, covering everything but production within the American Locomotive Company. When Walter Chrysler left the American Locomotive Company and joined General Motors Corporation in 1911 to become Works Manager of Buick, then headed by Charles Nash, Herman Weckler moved to Buick with him. Herman operated the Buick plant after Chrysler left General Motors to straighten out the Willys-Overland Corporation at Toledo, Ohio, and then the Willys Corporation at Elizabeth, New Jersey. Weckler continued operating the Buick plant until Chrysler asked him to come to Chrysler Corporation in 1932 to hire out to Keller, who had become vice president in charge of manufacturing.

The loyalty of our scattered engineering groups brought them back to our pre-war, centralized engineering.

Our pre-war styling had reverted from the advances of the Airflow to the boat bow pointed hood shape. This caused the hood lines to swing back, narrowing the windshield, and resulting in front body pillars that interfered with driver vision.

The styling of our post-war cars became higher, and rounded corners became more square. This seemed to be pushing our cars away from what the majority of the public was looking for as the seller's market gradually switched to a buyer's market. The public began to leave our cars for competitive models. Our competitors' cars at that time were lower, less boxy, their corners more rounded and further, they had curved windshields, something we had been proud of when we introduced them in 1934 on our CW Airflow custom cars. In spite of all the pressure brought to bear, our styling continued with the narrow front, flat glass, V-type windshields.

Skelton one day took it on his own to battle with President Keller to use curved glass windshields, but was turned down. When it became evident that the sales of our cars and profits had materially declined, the decision suddenly was made to get them into our cars as quickly as possible. This sudden decision to put curved glass windshields in all of our next year's models had a severe effect on the resale value of our former cars with their flat glass windshields which made them obsolete. It put us and our dealers in a difficult selling position.

This painful experience was followed by various attempts to split up our central engineering setup, the objective being to give each plant division its own engineering department. This to us was very disturbing. At the very inception of a Chrysler Corporation, we had gathered a close-knit team of engineers which had demonstrated the advantages of a centralized engineering operation. To this we had added a self-perpetuating feeder system of young engineers through our Chrysler Institute of Engineering. Nor did it hurt matters when our Chrysler Centralized Engineering was rated as a top example of efficiency. The enthusiastic acceptance of Chrysler Corporation's 1957 line of automobiles is proof of that excellence.

Conclusion

Reflecting on my many years of association with the automobile, after having retired some 25 years ago from corporation activity, I cannot help but make various personal observations that were unforeseen when I first entered the industrial world.

In my early years there were more than 2,000 makes of automobiles manufactured in the U.S. Most of these where conversions from the horse-drawn buggy days, and only a few ever lasted beyond their initial start of production.

The companies that initially prospered were designed by craftsmen. Flush with the surge of initial success, the top management officials inevitably would take over and decide that the next line should be enlarged, overriding the judgment that had led to their initial success. Such poor management decisions soon pulled down once well-established, reputable companies, which then had to borrow additional money and downgrade, ultimately going into receivership.

Fortunately Zeder, Skelton, and myself, with our diversified automotive background, recognized in starting our lifelong association in 1916 that to avoid failure, we should concentrate on laboratory and road testing, and provide accuracy in both our design and material specifications. We continually expanded our laboratory and research operations, forever exploring for new and greater results in development.

In looking back over the past, we can say that we must recognize that our universe is set up with innumerable fixed laws. These fundamental laws never change, are all related, and never inhibit one another. And they formed the basis for all of our engineering innovations at Chrysler.

Naturally corporations with good engineering foundations must be efficiently maintained, and conducted in balance with manufacturing, purchasing, sales, economics, public relations, etc. But they must also be run with the cooperation of a broad-visioned president and board of directors.

In concluding this biography, I cannot help but hearken back to the words of Herbert Hoover, once a renown graduate engineer, then president of the United States. Let me quote from his "The Engineering Profession."

> "It is a great profession. There is the fascination of watching a figment of the imagination emerge through the aid of science to a plan on paper. Then it moves to realization in stone or metal or energy. Then it brings jobs and homes to men. Then it elevates the standards of living and adds to the comforts of life. That is the engineer's high privilege.
>
> "The great liability of the engineer compared to men of other professions is that his works are out in the open where all can see them. His acts, step by step, are in hard substance. He cannot bury his mistakes in the grave like the doctors. He cannot argue them into thin air or blame the judge like the lawyers. He cannot, like the architects, cover his failure with trees and vines. He cannot, like the politicians, screen his shortcomings by blaming his opponents and hope the people will forget. The engineer simply cannot deny he did it. If his works do not work, he is damned.
>
> "On the other hand, unlike the doctor, his is not a life among the weak. Unlike the soldier, destruction is not his purpose. Unlike the lawyer, quarrels are not his daily bread. To the engineer falls the job of clothing the bare bones of science with life, comfort, and hope. No doubt as the years go by people forget which engineer did it, even if they ever knew. Or some politician puts his name on it. Or they credit it to some promoter who used other people's money. But the engineer himself looks back at the unending stream of goodness which flows from his successes with satisfactions that few professions may know. And the verdict of his fellow professionals is all the accolade he wants."

Looking back over a period of some 65 years, I cannot help but be impressed with the tremendous progress that the automotive industries have accomplished in a brief time, from my first ride in the steam car I built in a Los Angeles, California adobe building and drove for the first time in 1901, to 1916 when I joined with Zeder and Skelton who also had diversified horseless carriage automotive backgrounds, to our creation of Chrysler Corporation and the many new engineering innovations that we introduced to the entire automotive industry that have become accepted by all, until we finally retired in 1951.

Nothing would please me more today than to begin a tour from New York City to Los Angeles between two solid rows of Chrysler cars and trucks parked bumper-to-bumper on both sides of the highway. And as we rolled along, I would point out the many innovations we instituted such as hydraulic brakes, structural all-steel bodies, floating power, wide front seats and windshield vision, proper weight distribution with natural human comfort rhythm, and many more. Also, before arriving at Los Angeles, I would call your attention to the Airflow trend of styling I started in the 1934 Airflow series when it was so far ahead of its time with its wide, front, three-passenger seat.

I can only reflect with happiness on the part that Zeder, Skelton, and Breer played in the evolution of the horseless carriage.

Epilogue

Carl Breer maintained a remarkable collection of photographs and other memorabilia thoughout his lifetime. In going through his collection to select those that would conincide with the words of the manuscript, I could not help but put others aside simply because it seemed that they were too worthwhile to ignore even though I could not find a place within the text to which I could make them apply. I therefore have combined several of them as a closure to Breer's autobiography. Their captions will speak for themselves.

Anthony J. Yanik

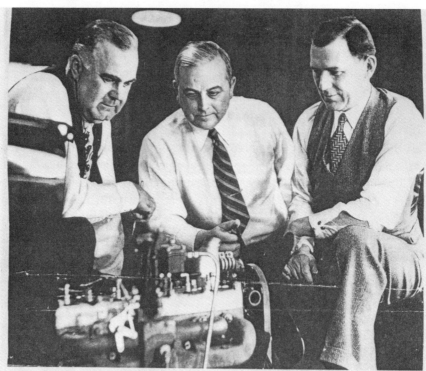

Throughout its earlier years Chrysler Corporation touted its engineering expertise, therefore it was not unusual for Zeder, Skelton, and Breer, preeminent as an automobile engineering team, to appear in its ads. This ad appeared in the Saturday Evening Post July 30, 1936 issue. (Breer Collection)

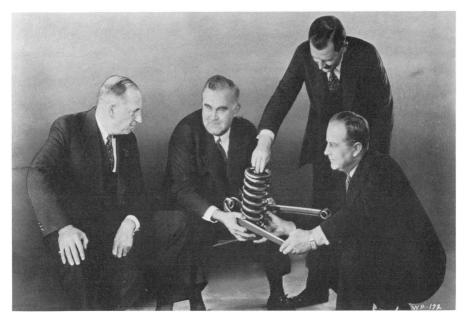

The Three Chrysler Musketeers demonstrate the advantages of coil spring suspension to Walter Chrysler. Coil spring front suspension was introduced on the 1934 Plymouth, Dodge and Chrysler Six. (Breer Collection)

The Chrysler Engineering management team in 1936. Pictured from left to right are Oliver Clark, George Allen, Carl Breer, Fred Zeder, Owen Skelton, and Harry Woolson. (Breer Collection)

The opening of the new Chrysler Engineering center in 1940 was the occasion of visits by many notable personages such as Dr. Compton, then president of M.I.T. (center). Carl Breer is at his right, Fred Zeder at his left. (Breer Collection)

In 1941, just prior to the entry of the United States into World War II, England's Duke of Windsor (center) received a tour of the Chrysler Engineering laboratories. Breer is to his right, Zeder to his left. (Breer Collection)

Carl Breer, as he appeared at his desk in Chrysler Engineering Headquarters on his 60th birthday, November 8, 1943. (Breer Collection)

Carl Breer, now in his retirement years, poses besides a restoration of his beloved Airflow, this one being a 1934 De Soto. (Breer Collection)

The Breer homestead on Windmille Pointe, Grosse Pointe, Michigan, about the year 1938. (Breer Collection)

The Birth of Chrysler Corporation

Carl Breer, Fred Zeder, and Owen Skelton in the twilight of their fabulous careers, December 16, 1947, at the road test headquarters at West Branch, Michigan. By 1951, Carl Breer had retired, and the triumvirate had been dismantled, but not before establishing its credentials as one of the most famous automotive engineering teams in automobile history. (Breer Collection)

On December 1, 1953, Carl Breer was honored with a Testimonial Dinner at the Detroit Golf Club in appreciation of his part in establishing Chrysler Corporation and giving impetus to its success. This is a photocopy of the program cover. (Breer Collection)

*For over 20 years friends and relatives of the Three Chrysler Musketeers —
Fred Zeder, Owen Skelton, and Carl Breer — would receive a Christmas card
featuring a sketch of the three on its face. Shown here is a montage of
20 years of those cards. (Breer Collection)*

Index

A
Air Temp, 89
Air trap, 132
Airflow car
    building of wind tunnel, 144-149
    coordinating body design with production, 149-150
    design guidance of nature's fundamental laws, 166-169
    developing first prototype, 151-163
    elimination of outside running board, 163-164
    eulogy for, 164-166
    exploring role of patterns, 143-144
    law of center of percussion, 169-175
Airflow principles, railroad ride research along, 177-185
Albert Champion Spark Plug Company, 97
Allis-Chalmers Manufacturing Company, 1, 5, 138
Allison Company, 55
All-steel bodies, change to, 114-117
Alternating current generators, 135
Aluminum Company of America, 94, 95, 96, 194
Aluminum pistons, development of, 94-96
American Can Company, 97
American Locomotive Company, 197
Amola Steel, 131
Amplex Corporation, 119
Associated Railroads, 183
Auto Car Corporation, 130
Automatic choke, 108-109
Automatic gear shifting, experimental studies in, 125
Automobile Manufacturers Association, 130

B

Ball, F. A., 44-46, 48, 108-109
Ball, F. H., 44-46
Ball, Tom, 48, 108-109
Ball and Ball carburetors, 44, 50
    two-stage, 63
"Belgian Rolls," 111-112
Big Six, 58
Biggs, Harry, 41, 57
Bofors mechanism, 191
Bohn Aluminum, 94-95
Borg Warner Corporation, 123
Borg Warner Overdrive, 123
Brakes
    four-wheel hydraulic, 79-84
    improving, 114
Breer, Bill, 11, 13-14, 18
Breer, Carl, xi-xiii, 199-201, 203, 204, 205, 206, 207, 208, 209, 211
    apprenticeship at Allis-Chalmers, 1-3
    building steam car in 1900s, 13-20
    and creation of Chrysler Corporation, 65-90
    decision to leave Willys-Overland, 70-75
    decision to rejoin Chrysler at Maxwell, 75-79
    early career of, 1-3, 9-29
    education of, 1
    emergence of four-wheel hydraulic brakes, 79-84
    and emergence of new Chrysler Six, 84-90
    family of, 9-10
    first job at Tourist Automobile Company, 20
    friendship with Skelton and Zeder, 5
    growing up in Los Angeles, 10-11
    roots of interest in engineering, 11-12
    at Stanford University, 21-29
    at Studebaker Research Engineering in 1916, 31-33
        curing crankshaft harshness noise, 40-41
        curing self-destruction of main bearings, 36-39
        early carburetor research and development, 44-51
        eliminating valve train noise, 39-40

Index   215

       fixing mainline engine, 34-36
       post-war Studebaker Engineering, 55-56
       redesign of 1918, 57-63
       research developments regarding universal joints, 41-43
       saving from receivership, 55-57
       searching for cause of high oil consumption, 36
       World War II days at, 51-55
    testimonial dinner for, 210
    at Throop Polytechnic, 20-21
    visit with Henry Ford, 135-137
Breer, Louis, 9
Briggs Manufacturing Company, 115, 161
Budd Manufacturing Company, 68, 115
Burgess, Magnus, 110
Burgess muffler, 126

C
California Institute of Technology, 1
Campbell-Wyant & Cannon Foundries, 114
Carburetor
    downdraft, 96
    early research into, 44-51
Caton, John, 138, 139
Caulkins, Bill, 118, 119
Centrifuse drums, 114
Chalmers plant, 77-78
Chilled cast iron valve tappet, 91-93
Choke, automatic, 108-109
Chrysler, Walter P., xi, xiii, 63, 124, 177, 178
    death of, 197-198
    and development of fatigue analysis, 110-111
    first contacts with, at Willys-Overland, 65-70
    visit with Henry Ford, 135-137
Chrysler CJ (1930), 98
Chrysler Corporation
    Air-Temp Division at, 192
    Amplex Division at, 120
    early product developments at, 91-142

all-glass sealed beam headlights, 127-131
aluminum pistons, 94-96
Amola Steel, 131
automatic choke, 108-109
breaking radiator monopoly, 98-99
change to all-steel bodies, 114-117
chilled cast iron valve tappet, 91-93
downdraft carburetor, 96
Electric Auto-Lite Corporation as spark plug source, 134-135
electric clocks and alternating current generators, 135-137
engine bearing battle, 99-100
engine exhaust muffling innovations, 126-127
ethyl gas, use at, 133-134
evolution of tires, wheels, and rims, 112-114, 127
experimental studies in automatic gear shifting, 125
fatigue analysis, 110-112
of floating power, 100-108
fluid drive, 124-125
fresh air heater, 131-133
helical gear transmissions, 120-121
hyraulic brakes, 84
improving brakes, 114
Keller overdrive development, 121-123
powder metal research, 118-120
revamping Maxwell Four, 93-94
rubber steering wheels, 109-110
vacuum tanks, 97-98
valve seat inserts, 117
visit with Henry Ford at Dearborn plant, 135-137
and World War II, 187-195
Chrysler Detroit Tank Arsenal, 190
Chrysler Four, 94, 105
Chrysler Institute of Engineering, evolution of, 137-142
Chrysler Royals, 115
Chrysler Six, 70, 84-90
Clark, Oliver, 63, 68, 116, 147, 148, 154, 163, 205
"Cle-Trac," 75
Cleveland Graphite Bronze Company, 99-100

Cleveland Tractor Co., 75
Clock, electric, 135
Clutch, Keller, 121
Coil spring front suspension, 205
Colby, Joseph M., 187, 190
Colt-Stewart, 115-116
Continental Motors, 73
Couture, Tobe, 69, 74, 76, 124, 133, 153, 190
Crankshaft harshness noise, curing, 40-41
Crankshaft impulse neutralizer, 100
Crankshaft torsional dampener, 100
"Cutless Bushing," 118
CW limousine, 156, 158, 159

D
DePalma, Ralph, 86
DeSeversky, Alex, 155
DeSoto Airflow, 157
Detroit Universal Products Corporation, 43
Devor, Don S., 65, 66
Diamant, Nick, 98-99
Diehl dynamometer, 33
Dietrich, Ray, 163-164
Dodge Four, 58
Doolittle, Jimmy, 155
Downdraft carburetor, 96
Dryden Rubber Company, 110
Duesenberg Motors Corporation, 66
Durant, W. F., 65, 73, 74
Durant, William C., 65, 71, 73-74
Durant Motor Corporation, 72
Duro Car Company, 1
Dynamometer checking tests, 67

E
Edsel, 136
Electric Auto-Lite Corporation, 134-135
Electric clock, 135

Engine bearings, 99-100
Ethyl Gas Corporation, 133, 134
Everett-Metzger-Flanders Company (E.M.F.), 5

F
Fairchild Airplane Company, 108
Fatigue analysis, 110-112
Fisher, Al, 174
Fisher, Fred J., 11-12, 13, 15
Fisher Body, 115
Fisher Brothers, 5, 116-117
Flint Car, 72
Floating Power, 100-108, 136
　　application to spring shackles, 107-108
Fluid drive, 124-125
Ford, Henry, 20, 115, 135
Ford, Henry, Company, 5

G
Gas turbine, 195
Gears, automatic shifting of, 125
General Aluminum and Brass Company, 37
General Electric, 138
　　Nela Park Lamp Laboratory, 128, 129
General Motors Corporation, 5, 130, 197
Generators, alternating current, 135
Goldman, Henry, 60
Good Maxwell Four, 105
Grey instrument, 179

H
Hall Lamp Company, 130
Headlights, development of all-glass sealed beam, 127-131
Heaslip, Jim, 51-52
Heater, introduction of fresh air, 131-133
Helical gear transmissions, 120-121
Hemingway, Ernest, 155
Herreshoff, A. G., 153, 162-163

Hodgkins, Rayal, 75
Holley Carburetor Company, 96
Home Electric Auto Works, 1
Hoover, Herbert, 200
Hudson Company, 130
Huessner, Dr. Carl, 150
Hump rim, 127
Hunting, 182

I

Imperial Airflow, 158
International Nickel Company, 74
Invar pistons, 95-96

J

Jamestown Metal Equipment Company, 99
Janeway, Robert N., 177, 178, 182
Jardine, Frank, 94
Jardine piston, 94
Johnson, Carl, 99-100
Johnson carburetors, 44

K

Keller, K. T., 121, 123, 138, 187, 189, 197-198
Keller clutch, 121
Keller overdrive development, 121-123
Kenney, George, 193-194
Knight sleeve valve engine patents, 65

L

Lancaster Dampener, 41
Law of Center of Percussion, 169-175, 178
Lee, Ken, 89, 90, 102-103, 125, 127, 128
L-head updraft carburetor, 73
L-head valves, 85
Liberty aviation engines, 51
Light Six car, 51, 58
Lionel, 177-178

Lobdel Manufacturing Company, 110
Lockheed, Malcolm, 80-81, 84
Lockheed Airplane Company, 79-80
Locomobile, 72
Los Angeles Water Works, 11

M
Manhattan Rubber Company, 81, 82
Martin, Royce, 52, 134-135, 136
Maxwell Four, revamping for power and endurance, 93-94
Maxwells, 88
Meinzinger, George, 58, 59
Mercury cars, 180
Moraine Products, 118
Moreland Distillate Truck Company, 101
Moto-Meter Company, 85
Motor Truck Company, 1
Motor Wheel company, 114
Muffler
    cutouts, 126
    developments in, 126-127

N
Nela Park Lamp Laboratory at General Electric, 128, 129
Nelson, Emil, 77, 95, 96
Newton, Isaac, 178
Northern, 1
Notch effect, 183

O
Oil by-pass filter system, 85
Oilite bearings, 118-120, 119
Oilless bushings, 118
Overdrive, development of, 121-123

P
Packard Motor Car Company, 7, 130
Parker Rust Proofing Company, 117

Index 221

Pelton water wheel, 11-12
Penberty Brass Foundry, 63
Percussion, law of center of, 169-175, 178
Perfect lip edge functioning, 82
Pfeiffer, Karl, 122-123
Pistons, development of aluminum, 94-96
Pope-Toledo plant, 7
Powder metal research, 118-120
Pressed Metal Products, 108

R
Radcliff transmission, 34-36
Radiators, 98-99
Rag joints, 58
Railroad ride research along airflow principles, 177-185
Rayfield carburetors, 44
Red Head engine, 134
Reed, Haines W., 21-22, 75
Regal Car Company, 5
Rims, evolution of, 112-114
Rollin car, 75
Rubber kick shackle, 107
Rubber mountings, 102
Rubber steering wheels, 109-110
Running board, elimination of outside, 163-164

S
Safety rim wheels, 127
Sauzedde, Claude, 77
Schebler air valve carburetors, 44, 46
Schwartzenberger, Frank, 74, 76, 84
Sheppard, Everett, 47-48, 96
Silent block bushings, 107
Sisson Company, 109
Skelton, Owen R., xi-xiii, 5, 57, 104, 112, 113, 124, 127, 153, 187, 190, 199-201, 204, 205, 209, 211
    and creation of Chrysler Corporation, 65-90
    decision to leave Willys-Overland, 70-75

decision to rejoin Chrysler at Maxwell, 75-79
early career of, 7
emergence of four-wheel hydraulic brakes, 79-84
new Chrysler Six emerges, 84-90
and transmission work, 120
visit with Henry Ford, 135-137
Skew gears, 120
Slack, Fred, 108, 190
Smith, A. A., Company, 60
Spalding, 1
Spark knock, 85
Spark, Whitigan Company, 97
Spicer shaft, 43
Sports and Classic Cars (Borgeson & Jaderquist), 165-166
Sprague dynamometer, 33
Spring shackles, application of floating power to, 107-108
Star cars, 73
Steering wheels, rubber, 109-110
Stout, Wesley B., 187
Stromberg carburetor, 51, 96
Stromberg Carburetor Company, 47
Studebaker, Clemente, 60
Studebaker Brothers, 5
Studebaker Research Engineering, xi, xii, 1, 31-33
    curing crankshaft harshness noise, 40-41
    curing self-destruction of main bearings, 36-39
    early carburetor research and development, 44-51
    eliminating valve train noise, 39-40
    M & S division at, 55-56
    post-war, 55-56
    pricing mainstream engine at, 34-36
    research developments regarding universal joints, 41-43
    saving, from receivership, 56-57
    searching for cause of high oil consumption, 36
    World War I days at, 51-55
Symington-Gould, 183

T

Throop Polytechnic Institute, 1, 20-21
Timken Roller Bearings Company, 180, 183
Tires
    development of, 127
    evolution of, 112-114
Toledo Steam Cars, 1
Tourist Automobile Company, 1, 20
Transmission, helical gear, 120-121
Trifon, 145, 151, 152

U

"U" shackle, 108
United Air Cleaner Company, 85
Universal joints, research developments regarding, 41-43
Universal Products Company, 43

V

Vacuum tank, 97-98
Valve seat inserts, 117
Valve tappets, 122
Valve train noise, elimination of, 39-40

W

W-125 models, 173
W-163 models, 173
W-196 models, 173
Ware, R. Dwight, 167, 169
Warner, Stewart, 97, 98
Waugh rubber cushion drawbars, 180
Weckler, Herman, 197
Westinghouse, 138
Wheels
    development of, 127
    evolution of, 112-114
White, Rollin, 18-19, 75
White Steamer, 1, 18, 75
Widman Body Manufacturing Company, 114

Wilcox-Rich Company, 91
Wills, Harold, 69, 128-129, 131
Wills St. Clair, 128
Willys, John N., 65, 66, 69
Willys Corporation, 70, 197
Willys-Overland, 7, 65, 70-75, 134
Wind tunnel, building, 144-149
WishBone springs, 66-67
Woodlite headlamp, 128
Woolson, Harry, 120, 139, 205
World War II, role of Chrysler Engineering Team in, 187-195
Wright, Wilbur, 144
Wright Aeronautical Engine Company, 188

Y
Young Radiator Company, 99

Z
Zeder, Fred M., xi-xiii, 1, 57, 92, 104, 110, 115-116, 124, 153, 163, 199-201, 204, 205, 206, 209, 211
    at Allis-Chalmers, 3
    and creation of Chrysler Corporation, 65-90
    decision to leave Willys-Overland, 70-75
    decision to rejoin Chrysler at Maxwell, 75-79
    early career of, 5
    emergence of four-wheel hydraulic brakes, 79-84
    friendship with Breer, 1, 3
    new Chrysler Six emerges, 84-90
    at Studebaker Research Engineering, 31-33
    visit with Henry Ford, 135-137
Zeder car, 75
Zenith carburetors, 44